Plant Breeding for the Home Gardener

Plant Breeding for the

Home Gardener

HOW TO CREATE UNIQUE VEGETABLES & FLOWERS

JOSEPH TYCHONIEVICH

TIMBER PRESS
PORTLAND • LONDON

Published in 2013 by Timber Press, Inc.

The Haseltine Building
133 S.W. Second Avenue, Suite 450
Portland, Oregon 97204-3527
timberpress.com

2 The Quadrant
135 Salusbury Road
London NW6 6RJ
timberpress.co.uk

ISBN-13: 978-1-60469-364-5

Printed in the United States of America
Book design by Marla Sidrow

Library of Congress Cataloging-in-Publication Data
Tychonievich, Joseph.

    Plant breeding for the home gardener : how to create
unique vegetables and flowers / Joseph Tychonievich. — 1st ed.
       p. cm.
    Includes bibliographical references and index.
    ISBN 978-1-60469-364-5

    1. Vegetables—Breeding. 2. Plants, Ornamental—Breeding.
    3. Vegetable gardening. 4. Flower gardening. I. Title.

    SB324.7.T93 2013
    635–dc23
                      2012023858

A catalog record for this book is also available
from the British Library.

To my very nearly superhuman
parents, for teaching me to love
learning. All the good things in my
life are pretty much your fault.

Contents

ACKNOWLEDGMENTS

Jackie Rosales, Kim Rupert, Andrew Barney, Adam Eckstein, and Sharon Kardos, for letting me share their perspectives on the joy of breeding plants.

Tom Fischer, for asking me to write this book and making it happen every step of the way.

The wonderful denizens of the Rose Hybridizers Association forum who, when I was a curious teenager with the breeding bug, were incredibly generous with seeds, pollen, advice, and information. It is no exaggeration to say I've learned more about breeding from you than from all my years in universities.

Kelly Norris, for fellowship in plant obsession and marathon idea-generating phone conversations.

My numerous friends in the gardening blogosphere. Without your support and camaraderie, I would never have considered writing about plants and gardening in any way. And now here is this *book*. Craziness.

INTRODUCTION

As a teenager I planted in my very first garden, among
other things, a few violas. Little miniature vigorous versions of pansies, which I absolutely adore. I put in a few
different varieties—black, purple and yellow, and pale
yellow. They flowered heavily for me in the spring, slowed
down and almost (but not quite) died in the heat of summer, then picked up again in the fall. I liked them, and I
wanted to grow them again, but the next year I forgot to
start the seeds. Come spring, I was kicking myself, saying
I had to remember for the following year, when I went out
to the garden to find it was covered with tiny viola seedlings. I had discovered the joys of self-sowing annuals.
More flowers, for no work at all!

When those seedlings started flowering, I was even
more delighted. In addition to the colors I'd planted the
year before, I had all sorts of new ones—dark purple with
a yellow heart, yellow with just a splash of lavender. Violas are notoriously promiscuous, and with the help of
bees, the different varieties had been getting busy, filling my garden with their incredibly beautiful love children. Once many of them were flowering, I spent some
very, very happy hours carefully examining them all,
picking my very favorite colors and combinations of colors to keep, and weeding out the others. The next year
they returned, in still more colors and forms, and again I
gleefully looked them over, comparing this against that,
deciding which ones I loved best. Every year, they came
back, and I liked them better, as the types I didn't prefer got removed, and the ones that were pretty to my eye
carried on. Not only were they more beautiful each year,
I noticed that each summer they seemed to get a little

better at surviving and flowering through the heat. They still preferred the cool of spring and fall, but without me doing anything at all, they were adapting to my particular climate, soil, and garden. One spring, while I was happily looking over the latest crop, the most beautiful, vigorous, and diverse yet, I suddenly realized something. These plants no longer looked or performed anything like those I had started with. They were, in fact, something new and unique, adapted to my tastes and my garden. My own personal strain of violas! How cool! Whatever should I name them?

I never did settle on a name (or if I did, I've forgotten it), and in the constant flux that was college, living in apartments and borrowing space to garden where I could, I lost the strain. But the lesson they taught me has stuck. Plant breeding is easy. Under the right conditions, in fact, you can hardly help but breed new varieties. Just nature and a little careful weeding will all but do it for you. The other thing I learned is that breeding your own plants is not just incredibly fun and satisfying, it is also practical. Not only was my strain beautiful, it grew better for me in my conditions than the commercial varieties I had started with.

I like to call varieties like that strain of violas "new heirlooms." Heirloom varieties are typically defined simply by their age, but I don't like that definition. It isn't the fact that they are old that makes heirloom tomatoes so delicious and heirloom dianthus so fragrant; rather, it is because most of these heirlooms came into being the same way my strain of violas did, from regular gardeners picking out what they liked best. While modern commercial varieties are bred with a least-common-denominator mentality, focusing on what will ship to and sell well at big box stores across the country, home gardeners focus on what they really love and get to partner with nature to create truly beautiful works of plant art.

As I became more seriously involved in gardening, many of the books I read dismissed creating new varieties as too complex to be attempted by anyone other than professionals. I knew this couldn't be true. My violas had proved that to me. More than a decade after those life-changing violas, I've bred dozens of different genera and studied plant breeding and genetics at two universities, and I'm still as convinced as ever by the lesson those little flowers taught me. Yes, it takes companies with teams of highly trained plant breeders years to create and market a new variety, but that is because the horticulture industry is complex, not because breeding is. Breeding at home is simple, easy, and effective.

Not to mention fun. Very fun. So welcome to my world. The world where every spring, I'm eagerly awaiting the first opening of a flower that no one has ever seen before, where I spend summer evenings with a friend and a pile of wonderful new tomatoes, trying each one, laughing over the odd-tasting misses and cheering for the ones that move me one step closer to that dreamed-of perfect tomato in my mind. It is the same process that brought us great tomatoes like the sublime 'Brandywine', an ancient art form that is long overdue for a revival. If you garden, you can breed plants. In this book, I'll tell you how.

From Teosinte to sh2

A BRIEF HISTORY OF PLANT BREEDING

1

've been growing a species of grass native to Mexico in my garden this year. To the casual observer, it looks like a typical ornamental grass: a clump of many slender stems with long, narrow leaves that move gracefully in the wind. With a little coaxing, it flowered for me in late summer. As with most grasses, the flowers are tiny—hardly recognizable as flowers at all. There is a little spray of male flowers, dripping pollen, at the tip of each stem, and tiny stalks of female flowers with long, silky stigmas growing at the base of some of the leaves. Later that summer the seeds matured, each stalk of female flowers producing a half-dozen small, glossy, incredibly hard, mottled gray and brown seeds.

This grass goes by the common name of teosinte, but to a scientist, it is *Zea mays*. I have another variety of *Z. mays* growing next to it, and this one looks completely different. Instead of many slender stems, it has one massive stalk as thick as my fist and taller than me, with hundreds of seeds, twice the size of those of the teosinte next to it, arranged on a long, thick, dense head. This plant, the same species as the grass beside it, is grown around the world; it goes by the common names of corn or maize.

I've grown corn, of course, many times, and I'd seen pictures of teosinte before, but actually seeing the two of them growing side by side, it is hard to believe they are the same species. Teosinte is, in fact, the wild plant corn was bred from. The change from wispy grass to the robust corn we all know is an incredible transformation, and I find myself in awe of the people who accomplished it.

AN EARFUL ABOUT CORN

Somewhere in what we would now call southern Mexico, early gardeners began one of the most amazing plant breeding success stories of all time, one that would eventually turn a small, scrawny grass with a few hard, nearly inedible seeds into corn, one of the most productive and useful plants humans have ever grown. The result of that prehistoric transformation now covers tens of millions of acres of farmland, providing a huge proportion of the calories that feed the world, and the basis of an ever-burgeoning industrial complex that takes corn and transforms it into everything from fuel for our cars to the "natural raspberry flavor" in our chewing gum.

The transformation from weedy grass to staple foodstuff was not quick. It took place step by step and was largely performed, until very recent history, by small-lot gardeners and subsistence farmers. Those early farmers and "breeders" worked on a simple basis. They planted seeds, looked at what grew, and saved seeds from things that they liked. They didn't—couldn't—simply come up with an idea and force teosinte to morph into what they wanted; rather, they had to sort through the options of new forms presented to them each year and choose among them. Teosinte itself was the chief collaborator in its own rebirth as corn, throwing out new versions of itself, mutations that humans could choose from. Over time, the seeds got bigger with softer, easier-to-cook seed coats. Ever-expanding numbers of seeds on each stalk resulted in that original tiny sprig of less than a dozen seeds becoming the big, bulky thing we know as an ear. And the many slender stems of the clumping grass were replaced by a single thick one, which allowed more plants to be packed into a smaller space.

Corn's many faces

Corn became the superstar it is because it is a uniquely mutable collaborator with gardeners. Not only did corn start producing enormous quantities of nutritious seeds for its human partners, it began reinventing itself in many other ways. Teosinte seeds are a dark, mottled brownish gray, but corn can be white, yellow, orange, red, green, blue, black—with multiple colors arranged in beautiful combinations along the length of a cob, and even striped or spotted individual kernels. Corn also transformed for multiple uses. Large, soft seeds, called flour corns, are easily ground into cornmeal; the harder flint corns are traditionally soaked and softened with lye before being ground and formed into tortillas and tamales. Some varieties are hard-shelled popping corns. Others—those with soft, sweet kernels when immature—are a fixture of summer picnics.

The corn revolution started in Mexico and slowly but steadily spread across the Americas. In its original wild forms, corn doesn't stop growing leaves and start growing tassels and silks until late fall or even winter. This is fine in semitropical climes, but in order to grow corn across the rest of the continent, it had to adjust. So as it spread, bit by bit, it adapted to mature earlier and earlier. By the time the Pilgrims arrived at Plymouth Rock, the local Indians had been all but wiped out by smallpox carried by earlier Europeans, but one of the greatest creations of native American civilizations, corn, stood tall; it was there to nurse the starving members of the colony through their first winter and went on to become a staple for the developing new country.

Going industrial

No doubt in the accidental devastation of European diseases, followed by the intentional devastation of European conquering, much diversity was lost, but America's corn was still a rainbow of colors and types. Gardeners and farmers continued doing what the native Americans before them had done, planting seeds, looking over their crops, and saving seeds from the most productive, beautiful, healthy plants. New types (like dent corn, a cross between traditional flour and flint corns) sprang up. These had higher yield but weren't as well adapted to the traditional cooking methods; increasingly they were used as animal feed rather than eaten directly.

Corn moved into the modern era with the advent of hybrid seed varieties (I'll explain, simply, how the supporting genetics work in chapter 4). Hybridization resulted in huge increases in vigor and yield, but also changed seeds from something farmers and gardeners save and select from each harvest to grow the next year, to proprietary varieties, carefully guarded, that companies produce and sell to farmers.

With the arrival of hybrid varieties, plant breeding—rather than being the domain of every gardener and farmer—become big business. Large companies formed around creating new varieties and poured huge amounts of money into the process. The intense focus and investment of corporate plant breeders has had a radical impact on corn. When hybrid corn varieties started hitting the market in the 1930s, corn yields in the United States averaged around 30 bushels per acre. In recent years, thanks to hybrid varieties and other modern farming techniques, yields have topped 160 bushels per acre. This has had a direct impact on our pocketbooks. In the 1930s, Americans spent an average of 24 percent of their income on food; today that figure is only ten percent.

Commercial breeding has very effectively taken corn a step further in its ever-changing identity. No longer a rainbow of colors and forms, each with a rich cultural heritage, it is a uniform, precisely engineered sea of yellow dent corn. Commercial corn breeders have accomplished their goals of high yields and maximum efficiency, but in addition to the debatable effect it has had on the health of our diet, the transformation has left the average gardener behind.

Rediscovering corn

A few years ago, I grew an heirloom flour corn variety in my garden, more out of curiosity than anything. The catalog claimed it had exceptional flavor, so I figured I'd give it a shot, but I was skeptical. Corn is corn, right? How different could the flavor be? I ground it into cornmeal and made a simple cornbread.

That cornbread was a revelation. I had no idea something I thought of as boring and a bit bland could be so delicious. But really, why should I have been surprised? The cornbread I'd been eating all my life was made from a plant bred to be high yielding and a good animal feed, with lots of fat to be extracted and sold as vegetable oil. Never once in those decades, among those crowds of highly educated corn geneticists poring over every detail of their crop, had anyone stopped to consider flavor. Corn, once the staff of life, has become an industrial feedstock.

Getting sweet

Even sweet corn has been turned on its head. Not too long ago, sweet corn was the ultimate fresh food. The moment you picked it, the sugars began converting into starches, so to really enjoy it, you had to grow it yourself or get it from a nearby farm, and all but run from the field to a pot of boiling water. Enter the genes *se* and *sh2*. Called "sugar enhanced" and "supersweet," respectively, these two gene variants showed up as random mutations, just like *su*, the original sweet corn gene discovered by native Americans; they make corn sweeter, but much more importantly they make corn that *stays* sweet during storage. This change makes it possible to ship fresh sweet corn all over the country and stack it high in grocery stores. Varieties with these genes are intensely sweet. They are also, well, a bit bland. I appreciated being able to buy these varieties when I was living in an apartment in a city and didn't have access to a garden, but now, well, I have a garden, and I want something a bit more flavorful and exciting. I want to grow something unique, something that you can't buy. The idea of sprinting across the yard with an armload of just-picked ears and plunging them into a pot of boiling water to preserve their perfect flavor is delightful, the essence of summer, in a way that pulling a bag of sweet but bland ears from the refrigerator just isn't.

Food for the soul

I also want my corn to be beautiful. If flavor hasn't been the focus of mainstream corn breeders, you can bet beauty never crossed their minds. I'm growing the variety 'Oaxacan Green' this year, and loving it. Each kernel is slightly different, ranging from bright emerald to a wonderful bluish turquoise, and the cornmeal I grind from them makes beautiful and delicious greeny gold polenta and cornbread. I also grow a variety of my own devising with dark red, almost black, stems that make a lovely contrast to the bright green leaves. The flavor is merely average, but it looks great growing at the back of my vegetable garden. It makes me smile.

ELSEWHERE IN THE GARDEN

The story of corn is far from unique; the same process has transformed almost every plant we grow. Move a plant from the wild into a garden, and over the generations, it begins changing and adapting to its new situation and the demands we make of it. Move plants from a small garden to an industrial farm, and they are reinvented again—but in ways that don't always make sense for the home gardener.

More isn't always better

Ever grow zucchini? Put in one plant, and you are deluged. The zucchini just don't stop coming. Anyone who grows it has a stockpile of endless, ever-more-bizarre recipes that attempt to use the bounty and jokes about neighbors hiding when they see a bag of excess produce headed their way. The one thing almost all gardeners can agree on is that the average zucchini plant simply produces too much. Talk to a zucchini breeder, however, and what is their top priority? Yield. As in *more* yield. They're actively trying to make that deluge even bigger.

I've grown zucchini for years, and every spring I'd plant it with a little hesitation, knowing full well I'd be sick of it before the summer was over. That all changed, however, the year I grew 'Costata Romanesco', a wonderful Italian heirloom variety. The plant was a little bigger, and when the harvest started coming, they were distinctively different. Each zucchini is long, dark green, with lighter green ridges that produce a charming star shape when you slice it. Beauty? Check! And flavor? Out of this world. The flesh is dense, firm, with a rich, satisfying nutty flavor, nothing like the bland, watery stuff I'd been used to. The yield? Significantly less, but I'm not complaining. A couple of plants still keep me eating zucchini constantly all summer and sharing a little with my friends, but not pushing me to the panicked point of digging up recipes for zucchini chutney. If I were a farmer, I'm sure I wouldn't grow it, because in that case, half the yield would mean half the money, but for my garden? It is perfect.

The connection here between flavor and yield isn't just chance. It takes energy for a plant to pump a fruit full of everything that makes it taste so delicious, so the easiest way to get a plant to make more and more zucchini is to have them instead fill each one with, well, water. So as long as grocery stores pay farmers by the pound, not per unit deliciousness, they'll keep demanding high-yielding varieties, and breeders will keep creating insipid, watery zucchini. I, on the other hand, have other priorities and will keep growing varieties like 'Costata Romanesco' that produce half as much but pack in twice the flavor—and are beautiful to boot.

Vanishing fragrance

Speaking of beauty, this story applies equally to flowers. In the quest for efficiency, much has been lost. Do an experiment. Head down to your local florist, and buy a bundle of carnations. Stick your nose in them and inhale deeply. Unless they've been sprayed with perfume (sadly, this isn't an impossibility), you won't smell much beyond maybe a vaguely musty, leafy smell. Now (and this will be the harder part) track down a plant of an old-fashioned garden carnation, or one of the closely related clove pinks. Get your hands on one of those, and there is no need to bend down and shove your nose in the flower— the rich, spicy scent will come out and meet you more than halfway.

The perfume that first convinced gardeners to welcome these plants into their gardens has been carefully and thoroughly removed from the modern florist's version of the carnation. Why removed? Well, cut your old-fashioned fragrant clove pink and put it in a vase with your modern, scentless ones, and you might see. Soon, the old-timey bloom will wilt and fade, while the modern ones carry on looking crisp and lovely week after week. Turns out that fragrance and vase life are inextricably linked, and when breeders worked to create longer lasting flowers, they also inadvertently removed their perfume.

For the commercial flower grower, a flower's ability to hold up postharvest is very important. Cut flowers, long before they arrive at your house, have been around the world. Grown mostly in Africa or South America, they are first shipped to the renowned wholesale flower markets in the Netherlands and then onto your local florist or grocery store. By the time you pick up that bundle of carnations, they may well have more frequent flier miles than you. In a market like that, tough, long-lasting flowers are a must. The more fragrant blooms, like the complex flavors of old-fashioned sweet corn, are a short-lived delicacy reserved for gardeners who grow them themselves.

The difference between what a gardener wants and what a professional breeder wants doesn't stop at the world of cut flowers. Petunias, like carnations, used to be scented, lightly during the day, powerfully so in the evening, the kind of sweet, evocative fragrance that wafts across the garden to you. Wander into a garden center full of petunias today, and you'll be lucky to smell more than a faint musty, sometimes even skunky smell. Just like carnations, we're faced with the fragrance-longevity connection. No scent means that each flower lasts longer, resulting in more flowers open on a plant at one time. I'm not always against that trade-off. In some parts of my garden, I'm willing to sacrifice fragrance for more flowers, but most of the time fragrance rules for me. There is nothing quite like sitting on the porch reading on a summer evening, enjoying the scents of my flowers moving on the warm air.

Making sales, not gardens

Fragrance is just the beginning. A few years ago, I had a chance to attend an event put on by the major plant breeding companies to show off their new varieties to the owners of garden centers and big box stores. As we walked around and saw each new introduction, I was stunned. I'd expected them to boast new colors or larger flowers, but those types of comments were actually rare. We'd see a display of a new series of pansies and learn that their great virtue is that all seven colors in the series come into flower in exactly the same amount of time after they are planted from seed, unlike the previous strains that had varied by a few days between different colors. Basically irrelevant to the gardener, this is critical for a wholesale grower so they

Kim Rupert breeds roses in Encino, California. When I asked her why she likes plant breeding, this is how she replied:

Until the question "What will those two make together?" stops popping into my head, I'll continue breeding. There are too many what-ifs that pop up every day, and it's *fun* trying to find out.

My seedlings will likely never earn me fame or fortune, but they *do* make my slice of dirt a happier place. My varieties are cleaner and healthier than many I've bought that did make it into commerce. It's a hobby, a pastime, and as such, it keeps me pleasantly occupied, mentally and physically.

Sharing your new rose is also a source of enjoyment. Just yesterday, my kid sister was deadheading a large, fragrant, thornless climber I raised. She had been working on it a little while when she paused, holding a yard-long spray of perfect flowers, and exclaimed, "I *love* this rose! It smells *so* good, and it doesn't bite! It's always in flower, and it's just flat-out gorgeous!" She went on to say, "I don't know all of your roses, but I'm sure as good as this one is, it will be considered one of your master-pieces!" That feels good!

What else is as easy, low-cost, fun, and exciting; teaches patience and permits sharing yourself, your imagination, your creativity; allows people to touch one another so sensually, intimately as well as across great distances? Creating your own plants does all that, and more, without requiring a great deal of time or skill, and provides so much more than it can ever require.

can plant everything at once, and then ship them all off to retailers in full flower on the same day.

Another common and even more disappointing boast was of new varieties that "don't require PGRs." PGRs? Most gardeners have never heard of them. PGR stands for "plant growth regulators," a group of plant hormone sprays used extensively in nurseries to keep plants from growing too tall. Short, compact plants are easier to sell because they look more balanced in a small pot. Short stems also don't get tangled together on the bench and are less likely to get broken during shipping; and you can stack many more flats of short plants on shelves in a semi-truck. So many, many of the plants (including vegetable transplants) you buy at a big box store or garden center have been treated with PGRs to keep them short and stocky. Theoretically, a plant will grow out of these treatments, so that once you put the plant in your garden, it can expand to its full height. Theoretically. I've certainly gotten some overtreated flats of annuals in the past that sat there, stunted and refusing to grow all summer.

Now, however, breeders are making PGRs less essential by breeding plants that stay short naturally. Short, compact plants aren't all bad; I like a few of them in the front of a bed, but I don't need a whole garden full of them. Snapdragons used to be glorious three-foot-tall spires. Now, good luck finding one even a third that height at a mainstream nursery. I've always loved the graceful growth of columbines—beautiful, delicate flowers on long stems dancing over the foliage. But they're impossible to ship without breaking those stems, so now we have little mounds, with flowers held right over the leaves. One shape of plant works best for the horticulture industry, so increasingly, just one shape of plant is all you can buy, even though every gardener knows you need a whole range of shapes, sizes, and textures to create a pleasing garden picture.

Gardeners choose plants for many different reasons. I want plants that grow and perform well without coddling. I want interesting, rich, exciting flavors, something I can't buy at a grocery store. I want beautiful vegetables and tall, elegant, fragrant flowers. Professional breeders and growers select plants on a whole separate set of criteria, ones that sometimes overlap in the areas of vigor and health but often don't connect with our more aesthetic interests. There's nothing wrong with efficiency, in its place. High-yielding varieties keep food prices low and fight world hunger, and cheap, mass-produced petunias make it that much easier for everyone to have a little beauty on their front porch. But when we gardeners put all the breeding of plants entirely in the hands of professionals, we lose a lot of the things we treasure most, the very reasons we enjoy gardening in the first place.

BY GARDENERS, FOR GARDENERS

In the world of tomatoes, some gardeners are taking control again. Like corn, tomatoes originated in South America. European explorers brought tomatoes back home with them, where they were, at first, greeted with suspicion due to their similarity to familiar toxic plants like deadly nightshade. Their amazing flavor quickly trumped those fears, perhaps in large part because tomatoes, almost uniquely in the world of vegetables, pack a punch of the so-called fifth taste, umami. Usually found in meats and cheeses, umami provides a rich, satisfying depth of flavor to foods, a welcome addition to the starchy bread and pasta staples of Europe at the time.

Tomatoes quickly conquered Europe and wormed their way into the heart of what became "traditional" cooking. Imagine, a moment, Italian food without the tomato.

As waves of European immigrants moved to the United States, they brought with them their favorite tomatoes from home. Though their time in the Old World had been brief, tomatoes had already started changing to suit local tastes and climates. You can still see dramatic differences in tomatoes, depending on where in Europe the variety comes from. Russian tomatoes are perhaps my favorites, varieties like 'Black Krim' which tend to have a rich, intense flavor and perform well in my cooler Michigan climate, while the Italian varieties I've grown have a brighter, more acidic flavor and a clear red color.

When these tomatoes arrived, they often didn't even have names, known only by the location where they originated or a person with which they were associated. Plants were simply passed down from generation to generation, changing slowly each year as gardeners saved seeds from what they liked best. The idea of a variety with a specific name and a very consistent set of characteristics is really a result of the modern world of commercial seed companies. Take the iconic tomato 'Brandywine', for example. Most people regard the original 'Brandywine', with pink fruits and potato leaves, as Sudduth's strain, from the Sudduth family in Memphis, Tennessee; but Amy Goldman, as she describes in her gorgeous book *The Heirloom Tomato*, discovered a much earlier catalog listing of a 'Brandywine' tomato with red fruits and regular leaves, from a gardener in Ohio. At this remove in time, it is impossible to trace exactly how these different varieties are related, but the probability is that just as the culture and climate of Russia produced numerous strong-flavored black tomatoes, Brandywine-type tomatoes developed bit by bit in the part of the country between Ohio and Tennessee as gardeners selected the individuals they liked best and shared seeds with family and friends. As commercial seed companies came into being, different strains were collected, named, and offered for sale. That amazing 'Brandywine' flavor, in other words, is the product of a community of gardeners, and the culture of a region, not just the work of one individual.

Old goals, new technology

Today, very excitingly, that kind of old-fashioned creation of exceptional varieties by a community to meet the needs of a community is coming back, thanks to the Internet.

On the online message boards of Tomatoville in 2005, Patrina Small happened to see a post by another member, Craig LeHoullier, about a few varieties of dwarf, indeterminate tomatoes. Unlike determinate types, these keep producing fruit consistently all summer long, and unlike typical indeterminates, they are small, compact, require minimal staking, and adapt well to containers. A pity, Craig said, that so few of these dwarf varieties are available, and that they don't have very good flavor.

That planted a seed in Patrina's mind, and that summer, she got busy and made some crosses between dwarf tomatoes and her favorite, most delicious and beautiful heirloom types. Inspired by the term "dwarf," she named her crosses after Snow White's seven dwarfs, making up new names (like Nosey and Pesty) when the number of crosses grew beyond seven.

Once you've made a cross, you mix up the genetics of the two parents, and the next step is to grow out lots and lots of seedlings so you can pick out the ones you like best. Keep doing that over five or six generations, and you'll get something that is stable and comes true from seed. That's what Patrina wanted to do, but how would she have the space to grow out the seedlings from all those crosses?

That's when she turned to the online community that had given her the idea, and the Cross Hemisphere Dwarf Tomato Project was born. They recruited other members of the Tomatoville forum to volunteer to take seeds, grow them out, pick their favorites, and send that seed back to Patrina and Craig. Working as a team, they could easily grow out large numbers of seedlings, the same tool commercial breeders use to identify the best varieties to meet their goals. Even better, Patrina lives in Australia, and Craig is in the United States, so they realized that by shipping seeds back and forth, they could grow two generations a year, one during the Northern Hemisphere summer while Australia was having winter, and the other in Australia's summer while much of the United States was under a thick layer of snow, allowing them to reach their breeding goals twice as fast. Many commercial breeding companies maintain winter nurseries in places like Costa Rica or Hawaii to do exactly the same thing, something that's usually out of the reach of average gardeners. By leveraging the power of their virtual friendship, Patrina and Craig were able to achieve the same effect for the price of a little postage.

The Dwarf Tomato Project introduced its first dwarf varieties in 2011, and several more are ready to be finished up and released. Traditional heirlooms are big and sprawling; they were selected by people with large gardens who would never have dreamed of growing a tomato in a pot on a patio. The project team members, a bunch of backyard enthusiasts, have created an incredible series of tomatoes that have all the traits gardeners really value: flavor, beauty, diversity—and all on plants that are uniquely adapted to the realities of small modern gardens. Wait for commercial breeders to create these varieties, and though you may have gotten the size, you'd never get the taste or eye appeal of these new tomatoes. The Tomatoville gardeners have created new heirlooms, carrying the traditions of generations of gardeners into the twenty-first century.

Rediscovering breeding

Stories like this get me excited. Whether it is just in your backyard, as a group with your local garden club, or with a circle of virtual friends scattered around the globe, you have the ability to take plant breeding back from corporations and create a new generation of amazing plants, new heirlooms that thrive in your climate, meet your needs, and are beautiful to you. The mechanics of the process are simple, comprising a few steps. First, you evaluate the diversity of plants available to you, getting to know them to understand what you like, what you don't like, and what you wish you could have. Next, you cross those plants to mix and recombine their traits, and finally you pick out from their descendants the individuals that you like the best. I'll explain each of those steps in greater detail in the coming chapters, along with some background information that will help you understand what is happening during each stage of the process, but really, almost all you need is an idea and the willingness to give it a try. The plants are still ready to change to meet our needs, and with a little creativity, you can participate in the grand collaboration between plants and people that has been going on since the domestication of plants began. So get out there, have some fun, and create something beautiful.

The Blank Canvas

MAKING BREEDING GOALS

2

I love zucchini. Over the years I've been gardening, I've tried many different varieties, but nothing I've ever grown can touch my hands-down favorite, 'Costata Romanesco'. It is uniquely beautiful and delicious, with ribbed, striped fruits and unsurpassed flavor. The yield is distinctly lower than most zucchini varieties, which actually isn't a bad thing. A plant produces enough that I enjoy it once or twice a week, but not so much that it starts driving me insane.

Summer squashes are generally well adapted to my zone 5 climate, enjoying our warm summers and regular rains. There is, however, a flaw in this practically perfect plant, a depressing reality that leaves me disappointed almost every year. Squash vine borers. If you are lucky enough to be unfamiliar with these little vermin, let me fill you in. The first sign of an infestation is usually when the plant collapses and dies virtually overnight. Everything is fine, lovely zucchini coming along, and then bam. It's all over.

THAT BLASTED BORER

The borer is actually quite a beautiful insect. If you have a bad infestation, you might see them in the early summer, vividly colored red and black with clear wings, looking, at first glance, more like a large wasp than the moth it actually is. Watch one, and you'll see her flit over a plant, settle down on the stem, and carefully lay an egg. Just that one little egg causes a big problem. Pretty soon, a little caterpillar hatches from that egg and starts eating. If they were content to just munch on leaves like most caterpillars, I wouldn't really care. A healthy squash plant can take a few holes in a leaf without missing a beat. Instead, out of sheer perverseness, they bore down into the stem itself, and start feeding there. Burrowing and munching their way through the stem, they quickly leave such a void that water can no longer travel from the roots to the leaves, and the plant dies as surely as if you had snipped the stem with a pair of pruners.

Since the borers live inside the stem of the plant, they are also a royal pain to control. You can't just squish them with your fingers, and even if I used insecticides (which I don't) it is hard to spray the inside of a stem. Recommendations for controlling this pest range from covering the stem with aluminum foil to carefully slitting the stem open and pulling the baby larvae out. Nothing I've tried has been particularly effective. One summer, my plan was to go out every single day, search the plants for the tiny, almost invisible eggs and try to squish them all before they could hatch. I suppose it could have worked, but life got in the way; I missed a few days, a few eggs managed to hatch, and the caterpillars slipped inside the stems and worked their evil magic.

In my last garden, the borer problem was so bad that I had to give up growing almost all my squash. I accepted that my zucchini crop would be delicious but short-lived, and limited my winter squash growing to butternut squashes. Butternuts are resistant to borers because they are a different species, *Cucurbita moschata*, whereas *C. pepo*, which includes zucchini and many other summer squashes, is basically candy to vine borers.

Trombone to the rescue

This depressing state of affairs continued for a few years until one winter I was browsing through some of my favorite seed catalogs, trying to forget the feet of snow outside, when a description leapt out at me. A summer squash called 'Tromboncino' described as being borer-resistant! In fact, this one was a variety of *Cucurbita moschata* like the butternut squashes I knew I could grow, unlike the regular *C. pepo* zucchini. Was this the solution to my zucchini woes?

That spring, I planted 'Tromboncino' next to my 'Costata Romanesco', and waited to see what happened. Once the two plants started growing, it was clear I was in for something quite different. While 'Costata Roma-nesco' formed a compact little bush and started pumping out delicious zucchini, 'Tromboncino' become a shock-ingly vigorous vine, quickly gobbling up an equally shocking amount of garden real estate. It began produc-ing later than my regular zucchini, but once the harvest arrived, I was quite taken. Smooth pale green skin, very long, thin, and twisted into bizarre shapes, and each with a charming little round bulb at the extreme end. They passed the pretty test, but the real question was what they tasted like. I harvested one zucchini from each variety and brought them into the kitchen. A quick sauté in olive oil and a little salt, and I was ready. 'Costata Romanesco' was everything it always was ... dense, firm, with a rich nutty flavor. Now for the 'Tromboncino' ... a decently firm texture, and some flavor. Better than the watery tasteless things they pass off as zucchini in the grocery store, but still nothing compared to my beloved 'Costata Romanesco'.

I was a little off 'Tromboncino' as this point, but just a few weeks later, as always, 'Costata Romanesco' collapsed and died under the attack of the squash vine borers, while 'Tromboncino' just kept on trucking, allowing me to harvest the entire summer, right along with my tomatoes. Too big, too late, and not quite delicious enough, but still, something is a whole lot better than nothing!

If only, I thought, I could combine that borer resistance with the flavor of 'Costata Romanesco' on a little more compact plant ... Wait, I can do this! Breeding project!

GETTING IDEAS

For me, breeding projects tend to start in moments like this, with me saying, "I wish that ..." or sometimes "I wonder what would happen if ... ?" I want my vegetables to taste better, my flowers bigger or smaller or a different color or more fragrant. I want more things to survive my cold winters, thrive in my summers, and I don't want anything to have problems with diseases and insects. So I start breeding to create plants that have those traits.

The process of starting a breeding project has two complementary, and deeply entangled, parts. I figure out what it is I *want* to create, and I search through the plants and varieties I can get a hold of to try to see what I'm going to be *able* to create. As with my zucchini example, those two parts often feed off each other, and a description in a catalog often leads to an idea for a breeding project. Though I'd long complained about borers, the idea of breeding a variety resistant to them never crossed my mind until I stumbled on the variety that might make that possible.

There are many different sorts of goals and approaches to designing a breeding project, and no one right way to do it, as long as you are having fun. But in my mind projects fall into four broad categories—different ways to look at things, to get inspired, which can help you find exciting ideas.

The best of both worlds

The squash project I just described is an example of what I call a best-of-both-worlds project. I have two varieties, each with a trait I want, flavor in one, borer resistance in the other, and I just need to combine them into a new variety. All I have to do is cross those two together to mix up their genes into new combinations, then pick out the individuals that combine the best traits of their two parents. These projects are great because you can be pretty confident they'll work out and give you something like what you've dreamed of. I know 'Tromboncino' has the genes for borer resistance, and 'Costata Romanesco' the genes for great flavor. It is just a matter of getting those genes together in one plant. This doesn't always work out, but a lot of the time it does. Many vegetables are just begging for similar breeding work. Older varieties often lack the vigor and health of modern breeding, while new varieties tends to fall down when it comes to fragrance and flavor, meaning there is a huge potential to cross between the two groups and have it all.

Andrew Barney lives in Loveland, Colorado. He started breeding a few years ago, inspired by a colorful ear of corn left over from a Thanksgiving decoration. The diversity caught his imagination, and he planted those seeds. He's now branched out to everything from beans to kiwis to squash, excited to see plants adapting to the harsh climate he gardens in.

My biggest accomplishment so far has probably been the moderate success of my watermelon land race. I tried mixing up as many different varieties of watermelon as I could find, planting them in a huge breeding mass. This is the first year I've even tried anything like this, but the results were pretty encouraging—first time I've been able to actually harvest a few ripe watermelons, and I have a huge amount of genetic diversity, which will help me to select plants that thrive in my environment.

Lots of things make gardening and plant breeding fun. I feel an almost spiritual connection to life, in many ways, when I'm outside in the garden. I guess it's a way to explore the universe: I don't have the ability to go exploring other planets, so instead I explore what's already here.

This approach has been used very successfully by a lot of talented breeders in the ornamental world. David Austin famously crossed antique roses with modern roses to combine the fragrance and form of the old with the rebloom and color range of the new. The modern incarnations of echinaceas, blooming in an increased range of vivid colors, are the result of a similar sort of cross made by Jim Ault, the brilliant breeder at the Chicago Botanic Garden. When he started working with the genus, all commercially available echinaceas were white or pink to magenta, but he had the inspiration to cross those standard forms with the very rare, and rather finicky, *Echinacea paradoxa*, the only species in the genus with yellow flowers. That cross let him combine the yellow tones with the pink, and the result has revolutionized the genus. Thanks to him, a stunning array of new yellow, orange, and red varieties have further inspired breeders around the world. Whenever you see a plant that is unique and different, that is an opportunity waiting to happen, a whole treasure chest of beautiful new plants to be explored.

Pushing past boundaries

Plants that already have the traits I want make breeding easy, but sometimes I want to push the boundaries, to create something that doesn't exist anywhere yet. This can be harder to do but is very rewarding. That's what I'm trying to do with artichokes. I live in Michigan, winter hardiness zone 5. Artichokes can only reliably survive winters in zone 7, maybe 6 in a sheltered spot. I love artichokes, both to eat and because they are strikingly beautiful plants. You can grow them as annuals in my climate, but convincing them to bloom and produce their first year is a rather involved process. Instead, I want fully winter hardy artichokes I can plant once and enjoy year after year after year. Problem is, hardy artichokes don't exist. There is no already hardy one I can simply cross with the delicious ones I want to grow. So here's what I'm going to try to do. From various sources, I've collected as many different artichoke varieties as I can. I've also got a whole list of cardoons, which are the same species as artichokes, just they've been bred for their edible leaf stalks rather than flower buds (sort of like cabbage versus broccoli). I've no reason to think they're any hardier than artichokes, but my goal here is just to collect as much genetic variation as possible. The more diversity I can shuffle together, the better my chances of getting a new combination of genes that will confer the trait, winter hardiness, I'm looking for.

I grew this whole range of plants from seed and planted them out in my garden. If I just put them all out in my yard, they'd almost certainly all die the first winter, and I wouldn't get anywhere. So, instead, I've planted them all against the south wall of my house. The heat leaking from the walls of the house, combined with the warmth of the sun, makes that the warmest, most sheltered spot in the garden. Warm enough that, some, but not all, of the varieties I've collected have survived. I'll interbreed all the surviving plants, grow their children, and let the cold of each winter pick the hardiest members of my population each year. Mild winters will allow more to pull through, while colder ones will knock them back to all but the very toughest. When I get a group of plant that can survive consistently in that sheltered, warm spot, I'll start planting them in the open garden, and continue crossing, planting seeds, and letting winter kill off the weaklings, until I've arrived (fingers crossed) at truly winter hardy artichokes.

The process here is the same whether you want to breed for winter hardiness, better flavor, bigger flowers, or some dramatic new color. The first step is to start with as diverse a group of plants as you can, so you have a lot of genetic variability to work with, and cross those all together. As their genes shuffle into new combinations each generation, carefully select the individuals that bring you closer to your goal, while maintaining as much diversity as possible to allow genes to continue to recombine and you to continue to make progress toward your goal. Cross the selected individuals together again, and keeping on repeating the process, year after year, until you reach your goal.

This type of slow and steady breeding work—unlike the dramatic results of, say, crossing a yellow echinacea with a purple one—is the main way most of our beloved plants were created. Early gardeners kept on saving seeds from *Brassica oleracea* individuals with the largest flower heads, developing larger and larger masses to create what we know as broccoli, while others saved seeds from the very same starting plant, but emphasizing the individuals with the tightest packed leaves, eventually creating cabbage. Gardeners selecting, step by step, individuals that improve on what they had before has transformed bearded irises from flowers with narrow, strap-like petals to modern varieties with wide, full, ruffled petals and all the wide variety found in modern roses and daylilies.

By its very nature, this type of breeding project is often never finished. Sure, someday (if I'm lucky) I'll be able to declare my artichokes fully hardy and stop breeding with them, but the reality is, I probably won't. Even once they're fully hardy, I'll continue sowing seeds, looking them over, and picking out the individuals that have the biggest or most delicious artichokes, or the most beautiful, finely cut leaves. If you grow plants from seed, keep them a little diverse by crossing things together, keep your eye on some goals of bigger, brighter, healthier, or more beautiful, then every year you'll be growing plants you like better than the ones the year before.

Getting crazy

There is a third sort of breeding project that I really enjoy, which I call getting crazy—shaking things up and seeing what happens. This is much less focused but a great deal of fun. An example would be my current (and ever-ballooning...) pepper project. It started when I was flipping through one of my favorite vegetable seed catalogs. The tomato section was *long*, with literally dozens of varieties, the flavor of each described in great detail with references to taste-test winners and the type of over-the-top adjectives you usually find in wine reviews. Then I flipped to the section with the sweet peppers and found ... not much. They were described as big, high-yielding, "pretty colors," and a couple merited adjectives like "delicious" or "sweet." That's it? No taste tests? Nothing with a "spicy flavor reminiscent of wild berries and musty leather" (okay, I made that up, but it isn't far from the real thing). Surely peppers could be just as exciting and interesting as tomatoes. Breeding project!

I don't really have a specific goal. I'm not aiming to create a pepper that tastes like any one specific thing. I just want to explore, to play, to see just how interesting and diverse sweet peppers can be. So I started collecting varieties of sweet peppers and making crosses to see what cool things might pop up, but really I was a little disappointed. I wanted crazy new flavors! Drama! And I wasn't seeing that in the descriptions of anything I was reading or in the plants in my garden. So I asked myself, what about other species of peppers? The sweet peppers, and most of the hot peppers we eat, are derived from the species *Capsicum annuum*. But I discovered there are a bunch of other edible species like *C. chinense* (habanero) and *C. frutescens* (tabasco). Most of these peppers are so hot that I've no idea what they really taste like under the heat. Maybe they could shake things up a little?

So I started collecting more peppers, and making crosses. The project is just getting started, but the results are already exciting. Exciting in more ways than one. Some of my crosses between a sweet pepper and a habanero are quite delicious and completely without heat. Others are so hot they bring tears to my eyes. Unfortunately, the only way to find out is to taste them. Russian roulette with peppers. It is quite fun though, and the results are intriguing. I'm already finding fascinating new flavors and starting to dream up new combinations of taste, texture, and size. As a bonus, the wide diversity of peppers in my garden means I'm finding plants that grow better and produce more than any other peppers I've grown—a happy and completely unexpected result that I'm thrilled about. That is often the way with breeding, and, indeed, life in general. You start with one goal in mind, but then nature presents you with something else entirely. So keep your eyes open, and don't be afraid to welcome the unexpected as it arrives.

The unexpected opportunity

The arrival of new, interesting and totally unanticipated traits from breeding projects is far from a rare occurrence. Echinacea, as I described earlier, was revolutionized by a cross that aimed at introducing a new flower color. Unexpectedly, some of the new hybrids also proved to be quite fragrant. It is a trait that hasn't been maintained in much of the new breeding, unfortunately, but who would have thought that crossing two minimally scented plants would result in fragrant offspring?

One of my favorite plant breeding stories is of a group of researchers who were working to create a more disease-resistant tomato. They found a wild species of tomato that was very resistant to disease. It doesn't look much like a tomato you'd recognize, producing tiny, hard, all-but-inedible green fruits. Not something you'd want to eat, but it had the trait they were after, resistance to disease, so they hybridized it with the delicious tomatoes we're all familiar with. As they reviewed the offspring from that cross, they found, as they'd hoped, individuals combining red, tasty fruits with the disease resistance they wanted, but they also found something surprising: individuals with deeper, more intensely red fruits than either of their parents. It turns out this tiny green-fruited species had a gene that, when combined with the right genes of the standard tomato, produces more of the red pigment, lycopene. No one, planning a breeding program for a deeper red tomato, would include a species with green fruits, but maybe you should. Bringing in more genetic diversity from wide crosses shakes up what is possible and opens new doors of opportunity.

✳

In the end, the process of coming up with goals for your breeding projects is something personal, spontaneous, and as hard to explain or teach as any other fundamentally creative project, but I hope the stories and examples I've shared here will help get your creative juices flowing. Start with those times you've said to yourself "If only!" and set about solving that problem. Keep your eyes open for varieties or species with a unique trait that can be combined with more standard varieties for an interesting effect. Keep selecting each year for the traits you like best, to keep on improving, and sometimes shake things up just to see what happens. Once you start thinking this way, you won't have any trouble coming up with new things to try. If you are anything like me, you'll instead be sitting around trying to figure out how to pack more fascinating projects into the space and time you have available!

The Birds, the Bees, and the Tweezers

HOW TO MAKE A CROSS

3

Whatever goals you settle on for your breeding projects, the basic skill you need to reach them is making crosses between different plants. In other words, controlling who has sex with whom. Despite the fact that flowers and humans are shaped rather differently, the fundamentals here aren't that dissimilar, and assuming your parents have had "the talk" with you, a lot of this should sound familiar.

FLOWER ANATOMY

Let's start with some basic anatomy. There is a lot of variation in plants, but most follow the same arrangement, as exemplified by a poppy flower. Around the outside is a ring of petals. These are, as you are well aware, usually large and colorful and often produce perfume and nectar to attract and bribe insects and other pollinators to visit the flower and carry the pollen to another plant. The size, shape, and coloration of petals vary wildly depending on how the flower goes about getting pollinated. Plants that rely on the wind to carry their pollen, like many grasses and trees, have very small or entirely missing petals. Plants like tomatoes that usually just self-pollinate, but rely on the occasional bee to come by and mix things up a bit, generally have small, fairly insignificant flowers. The bright, showy flowers we love in our ornamental gardens, on the other hand, are usually the hallmark of plants that depend on pollinators to get their business done.

Arranged, usually, in a ring within the petals are the stamens. Stamens are, in our little exercise in comparative anatomy, plant penises. At their tip, they have the anther sac, which is full of pollen, the plant version of semen. For most insect- and bird-pollinated plants, the pollen in the mature flower is exposed, fairly dense, and sticky, so it can be carried by the bee or hummingbird from one plant to another. Some plants tuck it away in special places, to protect it and keep it from getting eaten. Tomatoes, for example, keep their pollen rolled up inside the anther, where it can be slowly shaken out by the wind or the vibration of bees' wings when they land on the flower. Wind-pollinated plants produce huge quantities of very lightweight pollen that shakes out freely to drift on the air and with any luck find its way to a receptive

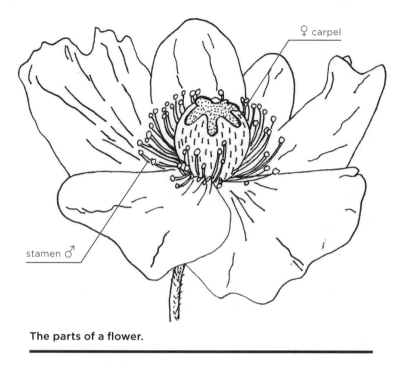

The parts of a flower.

female flower, sometimes making a stop to get inhaled by you and cause allergy attacks.

The target of all this pollen is almost always at the very center of the flower, the carpel, the female part of the flower, with the ovaries full of ova that will develop into fruit and seeds at the bottom and the stigma (the plant version of a vagina) at the tip, with the style in between. The tip of the stigma is covered with a sticky substance called stigmatic fluid. When a bee, or the wind, or you (after reading this book) place some pollen on the end of the stigma, the stigmatic fluid acts like glue to hold it in place; the fluid is also full of sugars to provide food for the pollen as it grows. Once in place, each grain of pollen starts growing what is called a pollen tube—a long

I know **Adam Eckstein** of Fort Collins, Colorado, through his work with roses, but his interest in breeding is wide-ranging. Like many breeders, he finds it brings rewards far beyond the beautiful plants he has created.

I cross just about anything at hand. It feeds my curiosity, and I feel a connection to the world in a way that Walt Whitman would understand. When I look at my plants, I know their very beginning. I know their early years. I recognize the stage in their life when they are still figuring out what they want to become. Finally at the end I can see what they have become, and it all humbles me and makes me feel a part of nature.

In this hobby you also get to meet some of the brightest, most generous, and most interesting people you will ever know in your life. We have people from all walks of life, all educational backgrounds, all sorts of different creeds—all working on doing something truly fascinating. I find myself marveling at the things I learn from them and at the things they have bred. We each have our own projects, but we are all fascinated enough with the whole thing that we are willing to help each other out. It is a beautiful collaboration of minds.

tunnel that pushes down through the tissue of the style to the waiting ova, where it fuses with the egg cells, to create a little baby plant—aka, a seed.

Now pregnant, the flower usually sheds its petals and other sexual equipment, and starts building seeds. The little embryo develops into a tiny plantlet ready to start growing when the time is right, packing in food in the form of starches, fats, and proteins to feed the seedling until it can get up and photosynthesizing.

MAKING A CROSS

Your job, as the breeder, is to interrupt the natural processes that carry pollen far and wide, and guide the pollen you want onto the stigma you are interested in fertilizing. In some cases, you can be pretty casual about this. I talked earlier about my first viola breeding. If you are mostly interesting in shaking up a few varieties and seeing what you can get, simply planting them together and letting the birds, bees, or wind do their thing can result in hybridization. This very casual sort of breeding can be fun and works well with promiscuous plants like violas and columbine, but it has its limitations. If you want to simultaneously breed both white and dark purple violas, for example, you'd be out of luck, because the two colors will keep cross breeding and producing various shades of lavender. Some plants, like tomatoes or beans, skip the whole bee thing, taking matters into their own hands and self-pollinating, meaning if you don't step in and make a cross, you could be waiting a long time for nature to do it for you. The simple task of making controlled crosses gives you the ability to guide the whole breeding process in the directions you are most interested in and to make crosses between plants that don't reliably interbreed; it also lets you work toward multiple goals at once.

Decoding different flowers

The first step in making a controlled cross is getting a flower and identifying all its parts. Almost everything you grow will follow the same basic plan as a poppy— petals on the outside, male bits in a ring, female parts in the very center—but there are lots of variations on that theme. When working with a new plant, take some time first to pull apart a few blooms, and make sure you can

Female (left) and male (right) holly flowers.

identify the different parts. While the basic model will cover most things, there are some oddball flower designs to look out for.

Squash plants, for example, don't have coed flowers; rather, they make separate male and female blooms on the same plant. For squash, the easiest way to tell the difference is looking for the little baby squash; you can see it at the base of the female flowers, but it is absent from the male. Corn actually does the same thing: male flowers we call tassels at the top of the plant, and female flowers (which will develop into the ears of corn we so like to eat) lower down. The silks on corn are actually the extremely long stigmas waiting for pollen from the tassels to land on them and grow equally long pollen tubes down to fertilize each individual kernel of corn.

A few plants (hollies and asparagus, for example) go one step further and are more similar to how we animals do it, with separate male and female plants, each only producing male or female flowers. Some—like *Arisaema*

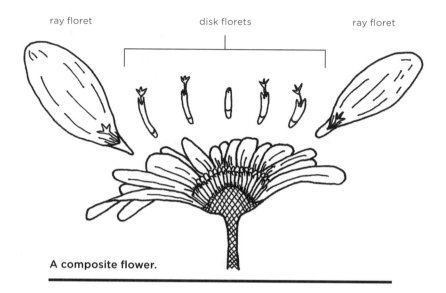

ray floret   disk florets   ray floret

A composite flower.

triphyllum, the common eastern U.S. wildflower known
as jack-in-the-pulpit—get even stranger: they start out
producing only male flowers, then switch to female after a
few years, and finally bisexual male and female blooms.

One confusing class of flowers is the daisy family.
Daisies, sunflowers, asters, zinnias, mums—they and
their kin all produce what are called composite flowers.
What looks like one large flower is actually dozens of tiny
individual blooms packed very closely together. What we
would typically call a daisy "petal" is actually one entire
flower called a ray floret, and the flat center of the bloom
is a whole mass of tiny disk florets. The structure of each
individual flower is actually the same as the standard
poppy, with petals on the outside—very, very tiny petals
in the case of the disk florets and a huge, lopsided petal in
the case of each ray floret—and anthers and stigma (or
sometimes just one or the other) in the center. Having
individual flowers that are so tiny and so closely packed
together can make them tricky to handle.

For most plants, it is easy to identify the parts of the flower. Look for powdery stuff, and you know you've got pollen. Look for a sticky bit, and you've found the stigma. If you get confused, a quick Internet search with the name of the plant paired with terms like "male," "female," or "dioecious" (the technical term for plants that have separate male and female individuals) will let you know if you need to be looking for separate male and female flowers, or if they have the more standard bisexual blooms.

Emasculation

Once you've identified your flower parts, the next step is to prevent pollen you don't want from getting in and screwing things up. Once a flower opens, it is designed to be pollinated, either by insects and other animals, the wind, or itself, so you need to get in there just before it actually opens to stop that from happening to ensure that the seeds you end up harvesting have the daddy you want them to have. There are two parts to this: preventing self-pollination by emasculating, and preventing cross-pollination by covering the flower.

To emasculate, you need to catch a flower bud just before it is going to open. Look around a plant and find the largest, most developed buds—these are the ones that are about to open and they are usually easy to find. The bud will be large, almost as large as the mature flower, and often beginning to color up. Take a pair of tweezers, pull the petals back, and carefully pull off the anthers. A small scissors, knife, razor blade, or (on the largest of flowers) your fingers can also be used, depending on the flower and your preference, but I find tweezers the easiest and most versatile. The flower has now been emasculated and cannot self-pollinate. For larger flowers, this is quite easy. For smaller blooms, it can be more difficult and require a

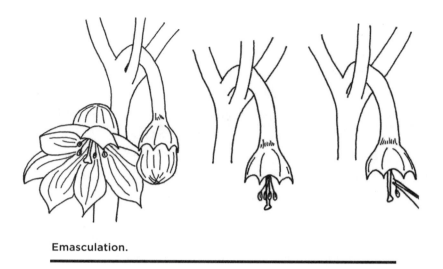

Emasculation.

steady hand. Be prepared to experiment with a few flowers at first to get comfortable with the process.

Keeping the bees out

Once you've emasculated, the next step is to prevent bees from coming along with pollen from another plant by covering the flower. I have used a couple layers of cheesecloth, nylon footie socks, or parchment paper held in place with a bit of wire, or a bit of aluminum foil. Breeders in dry climates recommend paper bags, but in my rainy summers they are a disaster. With some plants, with large, flexible flowers, you can use the petals of the flower itself to cover it. Squash is a perfect example of this. I simply tie a bit of string over the end of the developing flower bud; this holds it closed so no insects can get in. I've used this method for petunias, melons, and a couple others as well. If I ever take to breeding hibiscus, I bet it would work beautifully. You can be creative here, using any material that you have on hand that will keep bees from getting into the bloom.

. . . or not

Both emasculation and covering the bloom are often optional. A female squash flower has no male parts, and so obviously comes pre-emasculated, and some other plants are self-incompatible, meaning they won't set seeds with their own pollen. You can easily test if emasculation is required, by simply covering a flower without emasculating it, and see if it sets seeds by itself. If no seeds result, you are good to go, and can skip this step.

Covering the flower to keep pollinators out is also often not required. Since petals serve to attract pollinators to the bloom, if you completely remove the petals when emasculating, pollinators will usually not visit a flower at all, but that varies plant by plant, and region by region. In my garden, I rarely cover tomatoes and peppers and never have a problem. I always cover my squashes and hollyhocks, religiously, because bees seem to be addicted to them, petals or no. And wind-pollinated flowers, like corn, of course must always be carefully covered to exclude stray pollen. How careful you will have to be will depend on your climate, what species of bees you have around, and how important the cross is to you. If I am making lots and lots of pollinations and I'm willing to live with a little stray pollen, I'll skip it. When I'm just making one or two crosses, and I really care about the results, I tend to err on the side of caution. Again, you can easily test how important covering is in your circumstances. Remove the petals, emasculate, and then don't pollinate the flower. If seeds develop anyway, you know that something brought some pollen along. If the flowers drop off without producing seeds, you can be pretty confident you don't need to worry about covering the flowers.

Pollinating

Now that you've spayed the flower and fixed it up with a
little plant chastity belt, the next step is to collect pollen
from your other parent, and dab it on. The key is to look
for powdery (usually pale yellow, but not always) stuff on
the anthers. If the anthers look smooth and solid, than
they either aren't mature enough yet, or they're one of the
tricky ones I'll explain in a bit. The image most people
have in their minds here is collecting pollen with a paint-
brush, but I don't like that method. It isn't very exact,
and pollen tends to get stuck in the brush so you waste
a lot of it. The remaining pollen on the brush means you
also have to be very careful to clean the brush between
each cross, or you'll get all sorts of pollen everywhere,
sort of defeating the point of controlled crosses in the
first place. Depending on the plant, I have a few different approaches. If I have a lot of blooms to work with, I'll
just pick an entire flower, take the whole thing over to the
female, and rub its anthers over the stigma (sometimes
making bee buzzing or bow-chicka-wow-wow noises for
my own amusement). Sometimes I just take my trusty
tweezers and pull off an anther or two and take that over
to the female flower. For roses and other plants that tend
to be a bit stingy about producing pollen, I'll pick all
the anthers off and put them on a small plate to dry and
release their pollen; then I carefully dab that pollen onto
the stigmas, usually with my finger.

Between crosses, I clean the pollen off my fingers and
tweezers by rinsing them off with a little water or (if no
one is looking) simply a quick swipe with my tongue—
which really isn't all *that* gross, people, you eat pollen
every time you eat honey. Even if it is, erm, plant sperm.

For most plants, pollen is going to be easy to find as a powdery, sticky pile on the anthers. Some plants, however (tomatoes and peppers being the two you'll probably run into first), don't just shove their pollen out into the world; rather, they hold it inside the anther sac and wait for vibrations from the wind or arriving bees to shake it out. In these cases, you'll have to do a little work to extract the pollen so you can make your cross. It isn't too hard. Just hold the anther in one hand, and with the other, taking the point of a tweezer (or any sort of pointy object), split the anther open and scrape out some pollen. You'll be able to see it, a little pile of usually yellow or white powder. Take that powder, dab it onto your stigma, and you are good to go.

Wind-pollinated plants like corn won't usually have visible pollen either, because as soon as it is ready, it drops out and blows away. The traditional way to collect corn pollen is to pick the tassels and set them inside in a bag, where the pollen can be collected as it drops off; then that pollen is shaken over the silks. I've had good success with a much lazier method. I just pull off a chunk of tassel, set it on top of the silks, holding it in place with a bit of aluminum foil that simultaneously keeps other pollen out, while letting the tassel shed its pollen onto the silks.

Labeling

Once you've gotten the pollen on the stigma, your work is done. Just cover the flower up again (if you need to) and wait for the seeds to ripen. There is one critical last step, however. Label! It'll take time for your seeds to mature, and during that time, it is incredibly easy to lose track of which ones you pollinated. I think everyone uses a different labeling method, and it doesn't really matter what you do, so long as you always use it, you can find the labels, and once you label them, you know what they mean. I've settled on using bits of colored embroidery floss from the craft store, because it comes in so many colors. When I make a cross, I tie the string around the base of the flower, and then write down in my breeding notebook who the parents are. Breeding convention is that you write a cross like this: female (seed) parent × male (pollen) parent. So if I pollinated a flower on my plant of 'Black Krim' with pollen from 'Matt's Wild Cherry', I'll write: Black Krim × Matt's Wild Cherry. After that, I write "red" or "purple" or whatever color string I used to mark that particular cross. For plants with large, solid fruits like tomatoes and squash, I'll also, once the fruit is full-sized, write the cross in permanent marker on the surface of the fruit itself. That way even if the tomato drops off the plant or I harvest the squash by mistake and lose the labeling string, I'll still know what it is and not eat it by mistake.

Depending on your goals, you may want to pollinate just a few flowers or many. I'll discuss how many seeds you'll want for various purposes in the next few chapters, but how many flowers you need to pollinate to get that many seeds will vary widely. One petunia flower will produce hundreds of seeds, a single tomato or pepper will give you dozens, and a stingy plant like salvia makes only four seeds from each flower. Regardless, it is a good practice to pollinate some extras. Not every pollination will be successful, especially when you are getting started and are a bit clumsy with your tweezers, and I've had everything from deer, rabbits, and errant lawnmowers to my own carelessness eliminate developing seeds on plants. So even if just one pollination will give me all the seed I need, I always do at least two or three just to give me a little insurance.

Harvesting

After you've made the cross, the flower petals will drop off, and the fertilized flower will develop into a tomato or seed pod or squash. Wait until it is fully ripened and mature, harvest it, and take out your hybrid seeds. The specifics of harvesting seeds is a bit different for every plant but usually not too hard to figure out. If the plant produces a fruit, wait until it is fully or even slightly over-ripe, then squish the seeds out and clean off the pulp. For plants with pods or capsules, wait until they are brown and dry, then break open and separate the seeds from the pod. Some plants can be tricky because when their seeds are fully ripe, the pods will split open or even explode, spilling the seeds everywhere. Often these can be harvested just before they are ripe and be left to dry down the rest of the way in a paper bag, where the flying seeds will be contained. Alternatively, the developing pods can be covered with cloth or parchment paper to catch the seeds as they ripen. I give specific information for some plants in the final chapter of this book, but there are many excellent books on seed saving that give great detail for almost everything you could want to grow.

It is worth noting that when you pollinate a flower, the fruit that develops from it, be it a tomato or a squash or pea pod, won't be any different no matter what plant you pollinate it with. The seeds inside are hybrids, but the fruit is produced around the seed by the mother plant, and its looks and traits are determined by its genetics, not those of the seeds inside. Think of a pregnant woman's belly, which always looks the same, no matter whose baby she is carrying. The only exception is when you are harvesting not the fruit but the seeds themselves, and there are visual differences in the seeds of the different parents, the classic example of this being corn. The multicolored ears of Indian corn sold annually as fall decorations are patterned as they are because each kernel is a specific genetic individual, with different genes for various colors. If, for example, you pollinated a white corn with a mix of pollen from another white corn and a red variety, when you harvest the ear, you'll find a mix of red and white seeds, depending on which pollen got to which kernel.

Growing the babies

The final step of making a cross is to actually germinate and grow the seeds. For plants that are usually grown from seed, like most annuals and vegetables, this is a straightforward process, but for perennials that are rarely seed-propagated, it can be harder to track down that how-to information, and the germination process can be more involved. An Internet search pairing the name of the plant with the word "germination" will usually provide the information you need, but if not, you can figure it out. The most basic requirements for any seed to sprout is warmth and water. If that doesn't work, there are other factors to take into consideration.

Light. Some seeds require exposure to light to germinate, and others will not sprout unless they are in complete darkness. Usually, but not always, very tiny seeds will require light, while those demanding darkness are larger.

Cold. The seeds of many winter-hardy plants will not germinate until they have experienced several months of cold. This ensures that they germinate in spring, so they have a full season to grow before facing their first winter. To satisfy this stratification requirement, as it is sometimes called, sow your seeds in a pot or a moist paper towel in a plastic bag, and either put them outside over the winter or in the refrigerator for two or three months.

Scarification. Legume seeds, especially, can have very thick seed coats that prevent them from absorbing water and germinating. In nature, this is either broken down slowly in the soil by the action of fungi and other microorganisms, or it is softened by stomach acids as it passes through the digestive tract of a bird. You can—and I know a guy who does—feed your seeds to your pet bird, scrape up their poo, and sow it, but there are less disgusting methods. For large seeds, you can carefully abrade each with a nail file. For smaller seeds, the most effective method I've found is to put them in a bowl and pour hot, not-quite-boiling water over them, then let them soak in that water for a few hours. Successfully scarified seeds will start absorbing water and swell up quite dramatically (just like pre-soaking beans before cooking them). If some seeds don't swell, you can treat them again.

Smoke. Plants native to ecosystems that regularly burn often will not sprout without exposure to smoke. Luckily, you can find "liquid smoke" as a flavoring at your grocery store, usually next to the barbecue sauce. It is exactly what it sounds like, and watering the seeds with an added drop of liquid smoke will usually do the trick.

Freshness. Most seeds can sit dry, on your shelf, for months or even years, and still sprout just fine. Some seeds, called recalcitrant, cannot dry out or they will die; these need to be sown right away. On the flip side, a few other seeds have what is called an after-ripening period, and they won't germinate when perfectly fresh. However, most seeds germinate best when fresh, so usually planting them promptly will be your best bet.

Irregular germination. If you've only grown seeds from domesticated plants like vegetables and annual flowers, you're probably used to seeds germinating quickly and close to 100 percent. Wild plants just aren't like that. In the wild world, seeds germinate irregularly over a longer period of time as a kind of insurance against the entire next generation getting wiped out in one freak rainstorm or late freeze. So when you are working with a plant that hasn't been grown in gardens much, expect germination to be sporadic and spread out over a long time. If you need all the seeds to sprout, be patient. On the other hand, if you have more seeds than you can grow, if you only keep the seedlings that sprout the fastest and easiest, you'll be selecting for rapid germination, which, over time, will make your life easier.

SPECIAL CASES

The basic mechanics of cross-pollination are always the same, but there are some special cases where you might want to change your practices to have more success.

Difficult crosses

Most of the time, if you are crossing two different varieties of the same species, you should expect most of your pollinations to result in fertile seeds. But there are pairs of different species that can successfully but not easily make seeds. An example would be crossing a butternut squash (*Cucurbita moschata*) with a delicata or acorn (both *C. pepo*). These two species can interbreed, but only a small percentage of your attempts will be successful. The first thing to do in that case is simply to bump up your numbers: instead of pollinating two or three, keep on crossing as many flowers as you can until you see fruit starting to develop. If you can, also try the crosses with more than one cultivar, as often some varieties will cross more easily than others. It usually isn't hard to tell when you've had success—the petals of successfully pollinated flowers in most cases quickly drop off while the nascent seed pod or fruit begins to swell and develop. Unsuccessful crosses will generally unceremoniously drop off the plant. Frustratingly, some crosses will appear to be successful, but then abort, falling off the plant before they are mature. Roses are notorious for this. Again, numbers are your friend here, and the more you pollinate, the better your chances of getting mature seeds.

Reciprocal crosses

Since almost all plants are both male and female, you can make a hybrid between two plants two different ways, with either plant as the female parent. If you use plant A as the female parent in a cross and plant B as the male parent, then the reciprocal cross would be to use B as the female and A as the male. There are a few genes that are only inherited from the mother, but only a tiny percentage of the genome, so in most cases, it makes little difference which plant is used as the male and which as the female. But when trying to make difficult hybrids, often the cross will only be successful with one. So, if possible, try making both crosses and see which one works out better.

Bud pollination

Usually, the best time to pollinate is when the flower is fully mature and opened. But sometimes, pollinating just *before* the flower fully opens will give you more success. Some plants actually have systems in the flower that react with pollen, preventing some from growing and fertilizing the flower. But often these systems aren't up and running before the flower fully opens and can be circumvented by pulling the flower open and pollinating early. Many of the brassicas (broccoli, cabbage, kale, mustard greens, bok choy, etc.) react this way. If I'm trying a cross and having little luck, I generally will pollinate the flower a few days before it would normally open, when I emasculate it, and then keep on reapplying pollen every day until the flower fades. That way I'm fairly certain to hit the best time to pollinate, whenever that might be.

Time of day

Most flowers are receptive to pollen over a wide range of time, but a few are more picky. Usually, this is the case with flowers that are very short-lived. Squash, for example, opens new flowers every morning, and they then quickly fade by the afternoon, so pollinations need to be made first thing in the morning. Bean flowers also need to be pollinated in the morning. These plants are the exceptions, but it is worth keeping in mind when working with an unfamiliar plant: try crosses at several different times of the day and see what works best.

Storing pollen

For some plants, a barrier to making a cross can simply be that the two parents don't flower at the same time, making it kinda hard to get pollen from one flower to the other. In many cases, you can effectively save pollen from the first to bloom in the freezer and then use it to pollinate the second flower when it comes into bloom. The key here is to collect *lots* of pollen, because it is sure to be less viable after freezing than when it was fresh, and to make certain it is very dry before it goes into the freezer. If you live in a dry climate, simply setting it out for a few hours can be enough drying. I live in a perpetually moist part of the country, so I use a desiccant. Silica gel—familiar from those little "Do not eat!" packets you find when you buy shoes—sucks moisture out of the air. At craft stores, you can get big boxes of it for use in drying flowers. I take a bit of it, wrap it in a twist of paper towel, and put it with the pollen in a sealed container. I get very nice small containers for storing pollen from the craft store. They are intended, I believe, for storing beads, but they work beautifully for pollen.

I seal the pollen up with the silica gel, leave it at room temperature for a few hours to let moisture get drawn out of the pollen, then pop them in the freezer. When I'm ready to pollinate, I pull them out, let them warm up to room temperature without opening them (open the container while still cold, and water will condense on the pollen—and water is the death of pollen) and then dab it on my stigma of choice. This technique usually works, but to varying degrees. I've had rose pollen shipped to me from a friend on the other side of the country then stored it in the freezer for over a year and still had success using it. Corn pollen, on the other hand, is viable only when used very, very fresh. Trial and error is the only way I know to find out what can be frozen and what can't—just be sure to test the frozen pollen with a cross you already know can work with fresh pollen. If you try it first on a crazy cross you aren't sure about and it doesn't work, you won't know if it is due to the pollen being too old, or if those two species just won't hybridize.

✳

Making crosses can be tedious, but I always enjoy it. I love going out in the garden in the evening or first thing in the morning, looking over what is in bloom, and getting busy with my tweezers. Each cross takes less than a minute, and I'm done, left only to make a label, record the cross in a notebook, and wonder just how perfect and beautiful the babies will be when I get to see them.

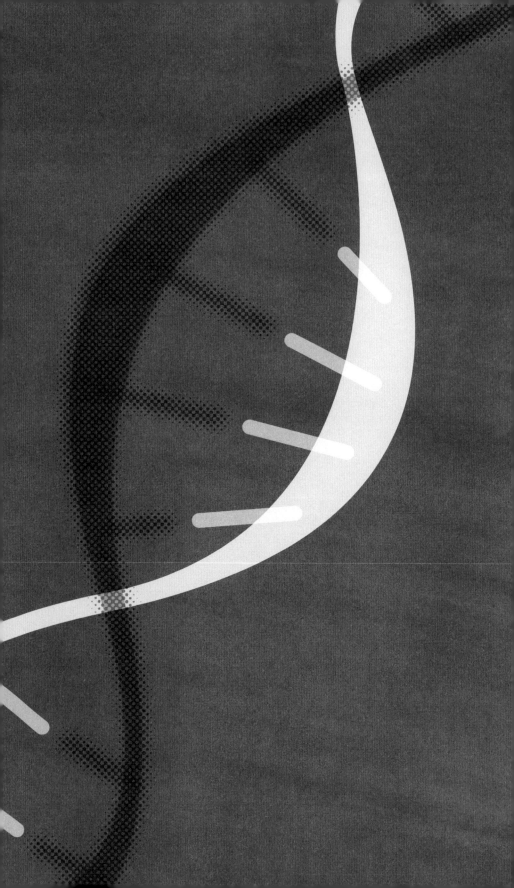

Genetics Made Easy

AND WHY IT MATTERS

4

You don't *have* to know any genetics at all to breed plants. You can just let bees and the wind make crosses and save seeds of what you like best, and you'll get interesting results. Early plant breeders worked that way, and they made a lot of amazing things. It is, however, a bit like trying to cook blindfolded, stumbling around adding things to the mixing bowl, hoping it turns out edible. A little information lets you see what you are doing and makes the whole process a lot easier and more fun. Genetics may sound intimidating, but the basic principles you'll be using in your breeding are pretty simple and actually quite familiar. My explanations here will just be to give a little background to tie what you already know from family photos together and translate it into how it applies to the plants in your garden.

GENES

Let's start with the real basics. Genes. They're little bits of DNA that are codes or templates for proteins. Proteins get made according to the sequence of the DNA, sort of like cookies from recipes or tools from blueprints, and then the proteins, which are essentially itsy bitsy little machines, go and do stuff. They can glom onto one chemical and switch it around into a new one, or carry it to a new location, or become part of a chloroplast and start turning light energy into sugar. In short, proteins make and control everything in a plant or animal or any other form of life, and genes determine which proteins get made.

Genes are often compared to recipes or blueprints, but, unlike buildings or cookies which are made from just one set of directions, most living things (including you) are made from two (or sometimes even more) sets of directions. You received one complete set of genes from your mother, and another complete set of genes from your father. You are the product of both of those sets. It is a little like having two recipes for cookies and following both of them when you cook. The first line of the recipe you got from your dad calls for flour, and so does the one from your mom, so you add flour. As you read through the recipes, everything is the same until you get to the final line, where the recipe from your mom calls for chocolate chips, while the recipe from your dad calls for oatmeal. You follow both directions, and hey presto, you get an oatmeal chocolate chip cookie! That's similar to how your genetic code works. The vast majority of the DNA is the same, and deals with the basic stuff of building cells and so forth, but a few genes, controlling things like eye color or how tall you are, are variable, and what you look like will be the product of the combined effects of the genes you inherit from each of your parents.

One gene or many

Now let's imagine you spill something on one of the copies of your recipe, and smear the "chocolate chip" part so much it is unreadable. That recipe has "mutated," and now that part of it doesn't work. So you just add what your other, unmutated, copy of the recipe calls for, and your oatmeal chocolate chip cookie becomes just oatmeal. If you had the two copies of the smeared recipe, you'd add nothing extra at all and get boring old sugar cookies. If the smeared section was for something more critical, like the instruction to place the cookies in the oven, things would be more dramatic. As long as you had one undamaged recipe, everything would be fine, but if both were missing the same part, you'd have a complete failure on your hands.

In genetics, this is how most recessive genes work—they're usually mutations of regular genes that no longer do anything at all. As long as you inherit a functional version of the gene from one of your parents, nothing changes, but get two broken copies, and the result is radically different. You probably learned in school about dominant and recessive genes because of dear old Gregor Mendel and his peas. He found a bunch of very distinctive traits in peas and figured out how they were inherited, thereby discovering some of the most fundamental genetic laws. The fact is, however, that these types of genes are actually pretty unusual in the real world. Just think about people. Things like albinism, dwarfism, whether your earlobes are attached, and blood type are controlled by simple recessive or dominant genes, but virtually everything else, from height to hair color to cancer risk to body type are not.

Essentially all the traits you actually care about are controlled by many different genes, and the variants are not the dramatically broken recessive genes, but simply slightly different versions of the same protein that work slightly different ways. To go back to the cookie recipe analogy, rather than one recipe calling for you to bake them and one not, different recipes vary in how long or at what temperature they are baked, producing a slightly moister or drier cookie depending on which you follow. Or, to think about it another way, though things like if a cookie has chocolate chips or not is controlled by a single line in the recipe telling you to stir them in, what makes a really great cookie is stuff like its texture, how chewy or crisp it is, and that trait results from the interaction of the relative amounts of sugar, fat, flour, and eggs you add, in addition to how long and at what temperature you bake them. There is no one single step in the recipe that makes them chewy or crunchy, and there is no one single gene in a person's DNA that makes them beautiful or ugly. Both are the result of many variable factors working together.

The same is true of plants. Most of the traits we care about are controlled by multiple genes, not one single dominant or recessive gene. However, for some plants, especially the more important food crops, some specific individually important genes have been identified and named. For example, determinate tomatoes have two copies of the recessive self-pruning gene, while indeterminate tomatoes have one or more copies of the dominant version of that gene. Most of the genes we know about and have names for are recessive/dominant pairs with dramatic effects that are easy to see, simply because the interacting effects of the many minor genes that control things like height and fruit size are harder to characterize. As a breeder, you don't actually really need to know much about these genes. If you know what genes control the traits you are interested, you'll be able—a little bit—to predict the results you'll get, but in the end, whether you know there is a gene called self-pruning or not, if you make a tomato cross, you'll be able to see which seedlings are determinate and which aren't by looking at how they grow, and pick the ones you prefer.

The biggest advantage to knowing about named genes is they can tell you how easy it is going to be to find individuals with the traits you want. Since determinate growth habit is controlled by a single gene, from any cross between the two growth habits, you can easily find individuals with either type of growth. But something like flower size, which is usually controlled by many genes, can be more difficult. Cross a small-flowered variety with a very large-flowered one, and you'll get a whole range of sizes in between, and it can often be hard to find anything either as small or as large as the parents. On the plus side, however, multi-gene traits give you more ability to fine-tune to your very favorite. A single major gene controlling flower size would give you only two sizes to choose from, while many genes mean you can pick from dozens of different sizes for what you like best.

Chromosomes and recombination

The many different genes in our genome are arranged into multiple chromosomes, which you can think of as multiple pages of a recipe. You have two copies each of 23 chromosomes, one copy inherited from each of your parents. Or, to go back to our recipe analogy, a 23-page recipe, with two copies of each page. When it comes time to make babies, you pass on just one copy of each chromosome, so imagine sorting through your recipe, and from each pair of pages, you randomly select just one to give to your child. On page one, perhaps the recipe from your father calls for whole wheat flour and brown sugar, while the recipe from your mom calls for white flour and sugar. You randomly pick the white flour–white sugar page from your mother, and go on to the next page, where your dad's recipe calls for oatmeal, while Mom's has chocolate chips. This time you might pick Dad's recipe, and so your child would inherit a white flour, white sugar, oatmeal recipe.

Because the directions for flour and sugar are next to each other on the same page, they are inherited together, while because the oatmeal/chocolate chip line is on a separate page, it is inherited independent of the flour directions. In genetic terms, this is called linkage. Genes that are located close together on the same chromosome are usually inherited together, while genes on different chromosomes are inherited independently of each other. Linkage is not absolute, though, and even genes on the same chromosome can be inherited separately, because of a process called crossing over. Imagine taking the two pages of the recipe, and cutting them in chunks, shuffling them, then taping them back together. If you happened to cut between the sugar and flour directions, you might get a new recipe page calling for brown sugar and white flour. This is what happens with your chromosomes. The two chromosomes pair up, and exchange bits, meaning that the closer two genes are on a chromosomes, the more likely they are to be inherited together, while genes which are far apart or on separate chromosomes will inherit independently.

MAKING BABIES: PATTERNS OF INHERITANCE

Let's go back to the cookie recipe analogy. You go through life, following the recipes you got from your

Jackie Rosales is a retired molecular biologist who has turned her scientist's mind to exploring the diversity and beauty of zinnias in her Indiana garden.

I've always liked working with plants—inherited the interest from both sets of grandparents. I just play around with easy plants, not for landscaping, but to see what all kinds of variations I can get. Mostly I have been generating a huge and varied pool of zinnias here in Indiana to see what sorts of variation and combinations of traits may occur in the offspring. Occasionally I'll select a new phenotype and see if a specific trait is inherited. The inheritance patterns have scarcely been determined yet at my place!

Zinnias are a natural choice, for me, since my grandmother always grew lots of them. They are easy to grow, extremely variable in appearance, relatively easy to cross, and can take the heat of our Midwest summers. And they attract butterflies, bees, hummingbirds, and goldfinches. For me, this is strictly a hobby! I always encourage people I meet to garden for the fun of seeing something being created as you watch it, for the beauty and color that it brings into your life. It need not be difficult!

parents, and then one day you get married, and start having kids. When it comes time to pass on the cookie recipes, you and your spouse can only give one copy each to the baby, so that it too gets one from Dad and one from Mom. Because you have two different versions, you randomly decide which one to give to your babies. Some get the chocolate chip version, some the oatmeal version. Each child will also get a recipe from your wife. Let's suppose that she too has one oatmeal and one chocolate chip cookie recipe. In that case, on average, one-fourth of your kids will end up with two oatmeal recipes, one-fourth will end up with two chocolate chip recipes, and half of them will have one of each, just like their parents.

You look nothing like your brother . . .

Because both of you have two different versions of the "chunky cookie mix-in" gene in your cookie recipes, your babies are going to be different. If your wife had dried cherry–macadamia nut cookie recipes, you'd get even more different possibilities, and none of your children would bake exactly the same sort of cookie as either of their parents do. Some would end up with a macadamia nut–oatmeal, which sounds kind of bland, while some lucky ones would be baking chocolate chip–macadamia nut (my favorite!). In stark contrast to the variability of those children, someone who has two copies of oatmeal, and has babies with another oatmeal-oatmeal cookie would have perfectly uniform little oatmeal cookie babies. In genetic terminology, if both copies of a gene are the same (oatmeal-oatmeal) they are called homozygous, while if they are different (oatmeal–chocolate chip) they are called heterozygous, and the differences between those two states has a big impact. If most of your genes are homozygous—both copies the same—you'll have children that look similar to each other, and look like you. If most of your genes are heterozygous—two different versions—your children will be extremely variable. To bring this back to the familiar, two Japanese people will have children that look Japanese. Two people of mixed Swedish and Japanese ancestry will have children with a much wider range of eye shape, height, and eye color.

The same thing happens with plants. When you buy a packet of seeds labeled "Brandywine tomato," all those seeds will grow up into plants that look essentially just like their parents, the plants who had sex to make those seeds. The variety 'Brandywine' is, in other words, a family of tomato plants which have mostly the same copies of their genes—they are highly homozygous—so that every time you grow the next generation (plant some seeds) you get plants that look and grow and taste essentially the same. The same is true of purebred (highly homozygous) dogs as opposed to mixed breed (highly heterozygous) mutts.

One of the big reasons this matters so much is that plant breeders (and dog breeders for that matter) need uniformity. People want to know what they are going to get when they plant a packet of seeds (or buy a purebred puppy). Commercial farmers *really* demand uniformity, because they need to have their whole field ready for harvest on the same day and in the same way. You can't be selling seeds that come up as a whole bunch of different-looking plants. People want 'Brandywine' to be 'Brandywine' every time they grow it. The most straightforward way to achieve uniformity is have highly homozygous plants, and the most straightforward way to get homozygosity is inbreeding.

Inbreeding

If you take a plant, and make it have sex with itself by self-pollinating it, then in the next generation, the number of genes which are heterozygous drops, on average, in half. Why in half? Well, think back to my example of the two oatmeal–chocolate chip people marrying each other. They are both—100 percent of that generation—heterozygous. But in the next generation, half their kids will either be homozygous oatmeal or homozygous chocolate chip, and only half their children will be heterozygous like their parents. All the homozygous offspring (if forced to inbreed, or "self-pollinate") will have all homozygous offspring, while the heterozygous ones will keep having only 50 percent heterozygous children. Keep inbreeding for a few generations and virtually all the genes will become homozygous, and each generation from there on out will look virtually exactly like its parents.

With animals, you can't usually have babies with yourself, but the same principle applies if you mated a dog with its litter-mate, or if you had babies with your sister or brother. It will take a few more generations, but you'll get to the same point. Basically, if you inbreed, you reduce the genetic variability with each successive generation. If, on the other hand, you let your kids run around with those dried cherry–macadamia nut people, variability will be maintained, and each generation will look different.

Now, I'm sure when I said "if you had babies with your sister or brother" you went, "*Ugh!*" because, on a gut, instinctual level, we know that inbreeding isn't a good idea. This isn't just a norm of human society; rather, it seems to be a nearly universal concept of anything that has sex, from dogs to plants to people.

Scientists call it inbreeding depression, or, on the flip side, hybrid vigor. Essentially, the more heterozygous (the more genes you have that are different) you are, the healthier and more vigorous you are (hybrid vigor) or, to say the same thing a different way, the more homozygous you are, the less healthy you are (inbreeding depression).

It is a pattern that has been found every time researchers have looked for it, though the size of the effect varies. For some species, like tomatoes for example, it is small enough to be hardly noticeable. For others, like corn and humans, it is so strong that inbreeding is almost lethal. Scientists have been busy trying to figure out what exactly causes this effect, and there are two main theories that, together, seem to explain most of it.

The first concept is "deleterious recessives." Let's go back to my example of the cookie recipe that mutated to lose the direction to bake the cookies. If this mutation happened to one of my recipes, I'll never even notice, because the other recipe still calls for baking, and everything is fine. I meet someone, have babies. Half of them get my broken recipe, but they're fine too, because they inherit an undamaged recipe from their mother. Even though this mutation is critical to producing edible cookies, it doesn't really affect anything, because to actually not get your cookies baked, you'd need to have the same error in *both* copies of the recipe. But that's only likely to happen if, well, we inbreed. As long as a mutation is rare enough, the only people in the world likely to have the same dangerous mutation I do are my immediate family. The reality is, mutations happen all the time, and we all almost certainly carry various strange, possibly lethal, recessive genes that no one knows about. Since your siblings are likely to have inherited some of the same screwed-up genes you have, if you marry your sister suddenly all those little errors in the recipe start getting put together and all sorts of problems can emerge. These problems make up a part of inbreeding depression. In other words, inbreeding can bring to light all the hidden deleterious or even lethal recessive genes that we don't know we have.

Another aspect seems to be that having two different versions of the same protein is often better than just having one. In terms of cookies, I guess this would be equivalent to a cookie with both oatmeal and chocolate chips being better than just one or the other. This isn't always true, but sometimes it is. In genetics, we call this effect overdominance. In this case we're not talking about broken versions of genes, but just different versions. Imagine a gene codes for a protein that performs some critical process of photosynthesis. There are two different versions, both of which work just fine, but perhaps one works a little better when it is hot, and the other works a little better when it is cold. A plant with both would be able to work well at a wider range of temperatures than a plant with just one or the other, and so be more adaptable and vigorous.

Outcrossers versus selfers

Whether the effect is due to overdominance or deleterious recessives, the bottom line is inbred = homozygous = not as vigorous. Most of the time. As I said, some plants are only slightly affected by inbreeding. The seriousness of the problem with inbreeding depends mostly on what a plant does in the wild. Many plants go to great lengths to avoid inbreeding, recruiting bees and hummingbirds to move their pollen over a wide area, timing the release of pollen before or after their female parts are receptive so they can't pollinate themselves, or even having systems to recognize and reject their own pollen. These plants are called outcrossers. Attempt to inbreed most natural outcrossers, and you've got problems. Corn, petunias, and people are all outcrossers.

Other plants take a different route. Depend on a bee to get your pollen around, and you risk not getting pollinated at all, so selfers take matters into their own hands (so to speak) and routinely self-pollinate. Adapted to this system for so long, these selfing plants show very little inbreeding depression, and include beans, wheat, and tomatoes. In other words, species that routinely inbreed have evolved over time to tolerate it. The bad recessive genes have been purged from their genome, and they do okay, while species that have adapted to outcrossing for a long time have lots of hidden recessives waiting to cause problems and have adapted to thrive with multiple versions of each gene. The only exception to this, that I know of, are the squashes, which, though naturally outcrossers, show very little inbreeding depression. No one knows for sure why, but it is possible that they've been inbred for so long in gardens that they've evolved to tolerate it.

Working around inbreeding

This business of hybrid vigor and inbreeding depression can become a headache for the plant breeder. As I said before, a variety needs to be uniform, and the best way to achieve uniformity is inbreeding. That works for selfing species, but for outcrossers, you can't do that without sacrificing the health of the plant. So how does a breeder achieve uniformity *and* vigor in the same variety?

The first, and oldest, method is a work-around, where you try to get homozygosity for the traits that matter to you, while leaving other genes are heterozygous as possible. This is what good dog breeders do. They look to mate two border collies that are as similar as possible in the traits (fur color, size, nose shape, etc.) that matter to them, but maintain genetic heterozygosity elsewhere by trying to breed with individual dogs that are unrelated. This is hard to do, and only works to some degree. Mutts almost always are healthier and live longer than a purebred dog, but a good breeder can, with time and patience, balance the two goals and produce healthy dogs that look the way they want. It is worth commenting that the more variability you are willing to tolerate, the better this works, and that does give an edge to the home gardener, who has more wiggle room than a commercial breeder who can only sell very uniform varieties. Especially in my flowers, I am happy to have each plant be a little different as it adds interest to the garden and keeps everything healthy and happy. It is also worth mentioning that the number of individuals you grow matters a lot here. If you grow only four corn plants each year, they will become inbred, simply because you can't avoid crossing them with a close relative. On the other extreme, grow a hundred of them every year, and you'll be able to maintain lots of good genetic diversity. I'll talk more about the specifics of maintaining healthy, outcrossing varieties later.

In plants, unlike dogs, you also have a couple of other options. Obviously, if you are working with self-pollinating crops, you can pretty much inbreed your heart out and have only minimal effects on the health of your plants. And it seems that early humans may have, without even realizing it, made that choice, because a disproportionate number of food plants are actually self-pollinating. Those first farmers surely didn't understand what was going on, but they must have realized that after a few years in the field, the self-pollinating species were doing better than whatever other species they may have tried to domesticate, leaving us with a bounty of inbreeding-tolerant plants for our gardens. Ornamental plants, on the other hand, tend to skew the other direction. Those large, showy flowers we like so much are the calling card of outcrossing species, because their purpose is to attract pollinators, something self-pollinating species don't need to do.

Cloning

Plants give you another out that doesn't work (at least not easily yet) with animals. Cloning. Oooh! Sounds all scary and science fictiony. Not really. Clones are all over the place—you've probably even met some. Clones are just individuals that are genetically identical. With people, you can (so far) get clones only when a tiny, fertilized egg splits in half, as happens naturally to produce identical twins. With plants, it is much easier. In some cases, very, very easy. Grab a potato. Cut it in half. Congratulations! You just cloned your spud: if you plant those two halves, each will grow into a genetically identical clone. Many plants are routinely cloned, including potatoes, roses, fruit trees, berries, and many ornamental perennials. Cloning neatly sidesteps the whole uniformity/inbreeding problem altogether. Just find a plant you like, divide it, take cuttings, or graft it, and hey presto, 100-percent genetically identical, perfectly uniform plants. If only we could grow new dogs from bits and pieces of an old one, fancy dog varieties could be just as healthy as mutts.

F_1 hybrids

So cloning and inbreeding-tolerant plants solve a lot of the problem, but what to do with something like corn? There is no economically practical way to clone corn, so you have to grow it from seeds, and inbreeding transforms it from a healthy, ten-foot plant to a scrawny little four-footer with tiny ears. Inbreeding depression is everywhere, but corn takes it to the extreme.

The answer is what we call F_1 hybrid cultivars, which just means the first generation produced from crossing two different homozygous lines. Remember the cookie recipe analogy? If I am a homozygous oatmeal cookie, I'll *always* pass an oatmeal gene on to my offspring. If I have babies with a homozygous chocolate chip cookie, she'll *always* pass on a chocolate chip gene. Which means, despite the fact that we are quite different, all our babies will be the same, they'll all get one oatmeal from me, and one chocolate chip from her. They'll all be the same— perfectly uniform—but they'll *also* not be homozygous. Because they are getting different genes from each of their parents, they'll be highly heterozygous, and therefore very vigorous and healthy.

That's what corn breeders do. They take two completely unrelated lines of corn, and inbreed them until they are very, very, homozygous. They then cross those two lines, and hey presto, the children are all the same, very uniform, but also highly heterozygous, so they are vigorous. Breeders call these F_1 hybrids (F_1 is short for Familial 1, and means first generation). To produce more of the hybrid variety, you have to keep on recrossing the two parental lines, because if you just save seeds from the F_1 hybrids, you'll get the next generation—called the F_2s—and those will all be different. Their parents are highly heterozygous, so they'll be very diverse, and they'll have on average twice as many homozygous genes, making them less vigorous. That's why you'll hear so often that you can't save seeds from a hybrid variety. You can, but it sort of defeats the purpose. The point of an F_1 hybrid is to be extremely uniform and also have a lot of hybrid vigor. Save seeds, and you lose not only the uniformity but some of the vigor.

There is an interesting twist to this, though. F_1 hybrid varieties are expensive and annoying to make. Instead of just selecting one ideal line, you have to breed and maintain two different parent lines and keep on crossing them to get your hybrid seed. For things like corn that show extreme inbreeding depression, it is very much worth it for the superior plants that result. But what about tomatoes? Tomato is a naturally self-pollinating species, with minimal inbreeding depression. There is little to be gained creating F_1 hybrids, and yet open a seed catalog and you'll see many, many hybrid tomato cultivars. In the case of tomatoes, it is simply a way to keep people coming back for more. The F_2 seedlings from an 'Early Girl' tomato won't be significantly less vigorous than their F_1 parent, but they won't all look or grow or taste like 'Early Girl'. Want that tomato? Gotta keep buying the seeds. You can, however, sidestep that, by selecting the best individuals from the F_2 plants, and inbreed them over several generations to create your own inbred "dehybridized" variety from the original F_1 variety. It won't be identical to the variety you started with, but you might actually like it better as you'll choose from the variability the particular traits that you personally like best.

So F_1 hybrid varieties are, in some cases, a stroke of genius. It is impossible to produce a nonhybrid corn variety that can match a hybrid for uniformity, yield, and vigor, and without hybrid seed technology, we wouldn't be able to produce nearly the amount of food from our farms that we do. For tomatoes, on the other hand, it is just a way to keep people buying seeds year after year. Some of you may be reading that, and thinking, "But the hybrid tomato varieties I've grown are much more vigorous and healthy than the nonhybrid heirloom types I've tried." This is true, but the difference there is due to the focus modern breeders have put on health and disease resistance, not hybrid vigor itself.

IN YOUR GARDEN

As a home breeder, you probably won't be making F_1 hybrid varieties. The space and effort required to keep growing out the parent lines and crossing them to get the hybrid seeds to grow simply makes it impractical. To keep outcrossers healthy, embrace diversity in your varieties, and grow large numbers. Remember, they don't all have to be in the garden at once. If you have room for only 20 corn plants each year, divide your seeds into three piles, and sow one set each year, saving seed each time. At the end of three years, mix all the seed you've saved together, and divide it into three groups again. That way, you'll keep mixing up the genes of all 60 individuals to maintain diversity in a smaller area.

If you are breeding with tomatoes or another self-pollinating species, you'll still be thinking about the F_1 and F_2 generations. If you cross two highly homozygous varieties—say, 'Brandywine' and 'Green Zebra'—the seeds you get will be the F_1 generation, which will be all the same. Self-pollinate that, and you get the F_2 generation which is where you'll see all the different combinations of the genes of the two parents. The very diversity that makes it so you "can't save seeds from hybrid varieties" is now your friend as a breeder, because this is your chance to pick out the individuals that combine the traits you like best—perhaps, a tomato that looks like 'Green Zebra' but tastes more like 'Brandywine', or a smaller-fruited, noncracking version of 'Brandywine'. If you keep self-pollinating your favorite individuals, each generation will show less and less variation. By the time you get to about the sixth generation, the plants will be almost completely homozygous, and come true from seed each year, like their parents did.

Parents F_1 F_2 ... F_6

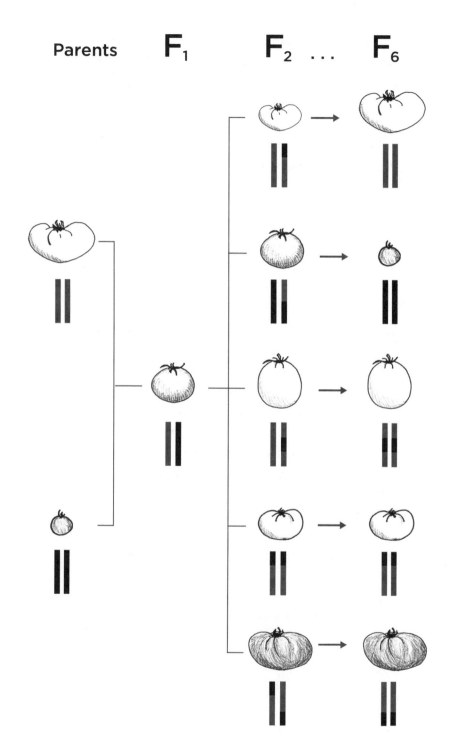

When you make crosses between clonally propagated plants (as many perennials are), the first-generation cross won't be uniform, because the parent plants aren't homozygous. Because there is so much diversity to select from in the first-generation crosses between clonally propagated plants like roses, many backyard breeders stop there; but you will miss a lot of variability if you don't grow out another generation. If you don't see the parental traits you admired in the first generation, it is still worth taking the time to cross some of your favorites from the first generation to see the increased variability of the next.

✳

That's it—the basic genetics that will help you understand what is happening when you are breeding and make good decisions about how to create the plants you really want. There are, of course, huge, ever-expanding bodies of information about genetics, as scientists dig ever deeper into the details of how DNA works, but despite having spent years studying hardcore genetics, virtually everything I do in my breeding work depends solely on the concepts I've just explained here. Armed with this same knowledge, you'll be able to understand what is happening when you go out into the garden and start creating beautiful and flavorful new plants.

Narrowing the Field

EVALUATING AND SELECTING YOUR FAVORITES

5

love every stage of breeding plants, but selection, picking out the individuals I like best, is my very favorite. When I make a cross, I've usually got something in my mind I'm imagining and hoping for, some specific flavor or color or growth habit. The reality when I actually get to taste and see the seedlings is consistently far more interesting and exciting than anything I could have dreamed up. Exploring the fascinating diversity is just great fun, with the added thrill of knowing I got to play a part in its creation.

Selection is also the part of the process where you really bring in your personal touch. As you can imagine, my garden is chock-full of my own breeding projects, and when gardening friends visit, we often walk around looking at and tasting my various populations. I'm always surprised by how different the responses are. I've literally had one person declare a tomato their very favorite while someone else spit it out because they hated it so much. I've walked friends up to a long bed full of different snapdragon seedlings, one of which I feel is *clearly* several times more beautiful than anything around it, only to see them gaping with amazement at something I was about to pull out because it was so boring and unattractive. If you took ten plant breeders, gave them the exact same lineup of seedlings, and set them to work, they'd all produce radically different final results because each would make such different initial selections. This is your chance to put your personal mark on the plants in your garden. It is also incredibly fun. One of my very favorite parts of summer is sitting with a friend and a table covered with tomatoes, happily munching away and comparing our impressions, dismissing the bland ones as "grocery store" tomatoes, and rivaling the excesses of wine critics in trying to describe the flavors of the ones we like best. I've also spent many a very happy afternoon crawling along a long bed of petunias, sniffing each one, jotting down its fragrance (or lack thereof), and pulling out the ugliest (I'm not a fan of magenta).

The basic idea here is simple: pick your favorites. But a little background information concerning special cases will help you make the most effective choices. So let's take a stroll through my garden together, and talk about some of my breeding projects and how I go about making my selections.

Evaluating the first generation

Walk out into my summer garden, and you'll see tomatoes. A *lot* of tomatoes. I'm a huge fan of them. They're delicious, easy to grow, and very fun to breed. I firmly believe you can't possibly have too many tomatoes. All summer long I gorge on them fresh from the garden, and when I have more than I can eat, which is most of the time, I cook them down into sauces or spice them up into salsa, and pack my freezer to the brim so I can feast on them all winter.

In one corner of my garden are a few short rows of three or four tomato plants each. These are my F_1 plants from crosses I made last year. As I explained in the last chapter, all members of the F_1 generation are essentially the same, so I grow only a few plants of each. I could get away with just one, but I grow extra as insurance against marauding deer or other disasters that can befall seedlings. One of the F_1s is a cross between two cherry tomatoes, one with terrific flavor but a horrible sprawling growth habit and very low yield which I bred myself (and call 'Wine') and the very compact, heavy-yielding 'Gold Nugget'. My goal in this cross is to combine the exceptional flavor of 'Wine' with the compact growth and high yield of 'Gold Nugget'. Since the F_1s are all the same, I'm not making selections, but I do take note of what the plants look like. Remember, the traits of the F_1 are going to be roughly the C-average of what I'll see next year in the F_2s. This particular F_1 is almost perfect—great flavor; loaded with fruit; short, compact plant. This tells me that next year I won't have to grow too many plants to meet my goal. I make a note to grow about a dozen F_2s next year.

Just next to those F_1s, however, is a cross between one of my all-time favorite varieties, 'Matt's Wild Cherry', and the best paste tomato I've ever grown, 'Opalka'. Here I'm aiming for a large, not-too-juicy paste tomato with some of the incredible flavor from 'Matt's Wild Cherry'. This one isn't looking so good. Flavor is so-so, and the fruits are tiny and all juice. To find what I want in the next generation, I'm going to have to be very lucky—and grow lots of plants ("Grow as many as possible!" is my note to myself). If I can, I'd like to grow as many as 50 or 60, but I might not have room for that.

Maximizing garden space

Walk further into the garden, and you'll see the F_2 tomato population I'm growing this year. This is another cherry tomato cross, aiming this time for a variety that will mature as early as possible. The parents are 'Matt's Wild Cherry' and 'Al-kufa'. Selecting tomatoes for earliness requires some care and planning because it can be strongly affected by the environment they are growing in. If one side of the bed gets more sun, it will warm up faster, and the plants on that side will bloom earlier not because they have superior genes, but because they are in a better spot. This is in contrast to traits like fruit color, and to a lesser extent, flavor, which aren't as affected by the environment. So for this population, I carefully selected a part of the garden that gets uniform amounts of sunlight and doesn't have low, wet spots or the super-fertile area where I spilled a bag of compost. In that carefully selected spot, I grew 40 seedlings, which is more than I really have room for, so I planted them only six inches apart. That's not enough space for them to mature, but since my primary goal here is earliness, that doesn't matter, because I'll be able to evaluate that trait while they are still quite young.

Early this summer, when they first started flowering, I watched their progress. The first 20 plants to flower, I kept. The other 20, who were later to bloom, got yanked, leaving more room for the remaining plants. Now that the plants are beginning to ripen their first fruits, I'm going through pulling plants that are ripening late and don't taste good, leaving plenty of room for the five or six earliest, best plants to mature, so I can see how well they grow and yield over the course of the season and make my final selection. By selecting as early as possible when the plants are small, I'm able to make the most of the space I have for growing seedlings to maturity.

Selecting early is a powerful tool to speed up your breeding progress and maximize your space, but it can also lead to problems. It is possible that one of those seedlings I yanked because it was a little later to bloom would have made up that time by ripening the fruit faster after flowering, or surprised me with stunningly delicious flavor that would make me forgive its lateness. On the other hand, if I'd grown only the six seedlings I have room to grow to maturity, chances are that exceptional seedling would never have gotten planted at all. In the perfect world of infinite time and space, I'd grow hundreds and evaluate them in every possible way. In reality, you need to focus on the traits you care most about and select for them as quickly as you can, so you can move on to the next stage of breeding.

The final steps

Beyond that bed is yet another group of tomatoes. Just ten seedlings this time, this is a breeding project in its final stages. These seedlings derive from a cross between 'Matt's Wild Cherry' (can you tell I'm obsessed with this variety?) and 'Black Krim'. This was a project I worked on with Lynne Rossetto Kasper, host of public radio's food show *The Splendid Table*. I grew the F_2 population and shipped her the fruits; she tasted them all and picked out the one she liked best. The perfect tomato singled out, I'm now in the process of inbreeding it. Last year I grew about 20 of the F_3 plants, picked the one that matched the F_2 plant Lynne had selected, and this year I'm growing the F_4 plants from that. With each generation of inbreeding, they show less and less variability, so I can grow fewer and fewer plants. In another year or two, the variety should be stable, all the seedlings looking the same, and I'll be ready to start sharing it. Next to those plants, I have a few plants that aren't breeding projects but simply mainstream hybrid and heirloom cherry varieties I'm growing alongside, so I can compare their growth and flavor to my selections. I like to do this whenever I have the space in a breeding project; it gives me a better sense of whether the plant I've developed is really superior to or different from what is already available. I'll enjoy them whether they are or not, but direct, side-by-side comparison helps me decide if a variety will just stay in my garden, or if I'll be sharing it far and wide.

In each stage of my tomato selection, I pick just a single plant to harvest seeds from and carry on to the next generation: that is the quickest way to a stable, uniform variety since tomatoes, as a self-pollinating species, can take the inbreeding. I look carefully at the F_1s to decide how many F_2 plants to grow, and make selections as early as possible in the F_2 generation to save on garden space. When it comes time to select, I prioritize the traits most important to me, be it earliness or flavor, and evaluate those first, only afterward looking for things I don't personally care as much about, like color and shape. As I see variability go down in subsequent generations, I grow fewer and fewer plants each year to make room for all the other projects I have dreamed up.

Selecting outcrossers
Wander over to the other side of the garden, and you'll see my corn. Selecting corn, an outcrosser with strong inbreeding depression, is different from selecting tomatoes, which can be inbred with little problem. Instead of going through and picking out the one perfect individual,

I have to select a group of individuals that get me closer to my goal but still maintain some genetic diversity. This stand of corn is part of an ongoing project to breed the biggest possible popcorn kernels. Every year I plant about 100 seeds. At the end of the season, I harvest, putting the ears from each plant in a separate, labeled paper bag. Then I go through each bag. I pick out individuals with the top ten biggest kernels, and set them aside. Then I go through again, and try popping a few kernels from each ear, and pick the ten that pop the best. The ten biggest kernels and those ten best poppers provide the seeds I'll plant next year to create the next population. Every year I select 20 of the 100 plants to continue the project.

I picked those numbers—20 of about 100 plants—from a very cool research study that has been going on at the University of Illinois. They've been doing exactly that, picking the top 20 of about 100 corn plants for oil content and protein content, and using that selected 20 percent as the seeds for the next year; and they've seen steady improvement in the selected traits each year—for over a century now. That tells me that 20 plants maintains enough diversity to keep corn improving and avoid serious inbreeding depression. If I had the time and space, and could grow and evaluate thousands of plants, I'd reach my goals faster and have less inbreeding depression. But I can't, and so I use 20 plants as the minimal number to select to maintain diversity in my corn (and 20 percent as a rule of thumb with other outcrossing species). If I didn't have room for 100 plants, I might break things up over different years, by splitting my selected seeds into two piles, and sowing 50 of them one year and 50 of them the next. Then I'd mix together selected seeds from those two years and use them to plant for another two years. That way I could maintain genetic diversity but in half the garden space.

One final way that selecting an outcrosser like corn is different from a selfer like tomatoes is dealing with recessive genes. For example, suppose you crossed a blue corn with a white corn and want a pure blue variety in the end. Blue is dominant, so only kernels with two copies of the white gene will be white (the blue kernels will either have two copies of the blue gene, or one blue and one white). If you just select out all the blue kernels, you'll get mostly blue in the next generation, but because the white color is recessive it "hides" behind the blue gene, and some of those blue kernels will have white babies. There is no way to tell just by looking which kernels are all blue, and which are blue-white. If you just keep on selecting against the white kernels, their numbers will go down in the population, but in the large, diverse populations required to keep corn healthy, they'll never entirely disappear, and whites will keep popping up every now and again, ruining your uniformly blue corn.

The fix is pretty simple. All you need to do is self-pollinate your population for one year. When self-pollinated, the blue-blue individuals will produce all blue seeds, and the blue-whites will produce roughly one-quarter whites, three-quarters blue. For corn, this is especially easy, because each individual seed shows the phenotype of the next generation, so you can pick out the blue-blue individuals just by looking at the cobs. For any trait that isn't visible in seeds—which is virtually everything you care about in almost any plant other than corn: flowers, leaves, fruits—you'll have to plant out the seeds from each individual in separate blocks so you can observe which ones are free of the recessive gene. This one generation of selfing will introduce a little inbreeding depression, but as soon as you mix all your individuals back together and let them interbreed again the following year, their different genes will get recombined, restoring their hybrid vigor.

Selecting self-sowers

For outcrossers like corn, big populations and careful selection that maintains diversity are the key. Come up closer to the house, to my shade garden beds, and you'll see I'm doing the same thing, albeit in a more lackadaisical fashion, with my columbines. Columbines, like corn, are confirmed outcrossers and promiscuous to boot, and, with the help of insect pollinators, they spread their pollen far and wide. They also happily self-sow, so breeding them is almost effortless. I've simply collected forms and colors I love, with a big emphasis on fragrant varieties, and let them do their thing. Early summer, when they are in peak bloom, I walk through the beds and look them over. I like my columbines very tall, with long, long, graceful spurs, strongly fragrant, and vividly colored. Individuals that meet all those requirements I let set seed to carry on their genes to the next generation. If a particular, truly exceptional plant catches my eye and nose, I'll harvest its seeds, and start them indoors to beef up the numbers of its children in next year's garden. Plants that don't quite meet those requirements but are otherwise interesting—maybe they have weak color or short spurs but exceptional fragrance—I let bloom, but I deadhead them before they set seed to limit their impact on the next generation. Anything I really don't like—namely doubles, or fat dumpy flowers—gets weeded out completely to ensure their genes go no further.

The final result is basically the same as what I do with my corn, but I'm less careful about it. No careful counting to make sure I have about a hundred plants, or calculating what percent of the population I'm selecting. This is just a personal choice, and it may change. I mostly love my columbines as they already are, so I enjoy gently, casually nudging them in the direction I want them to go. If, some day (as usually happens with me . . .) I get a bee in my bonnet about some new direction I want to take my columbines, I'll probably get more careful, making controlled crosses and rigorously evaluating them. For now, the casual style is working well for me, and every year they are more beautiful.

Selecting perennials

Back in the full-sun part of my garden, next to the tomato beds and scattered here and there where I have room, are my rose seedlings. Unlike corn and tomatoes, which I grow from seed each year, roses are perennial and can be propagated by cuttings. I neither have to inbreed nor try to maintain genetic diversity: when I find the perfect individual, I can just clone it by taking cuttings. I approach selecting roses a little differently than I do many other plants, because there are so many different traits I am interested in. Disease resistance, color, fragrance, form, rebloom, flower size, foliage color, shape of the plant, and winter hardiness, to name a few. To help me take all those disparate traits into account, I take a list with me out to the garden and give each plant a score from 1 to 10 on each trait. Taking the time to look closely at each attribute helps me notice the weaknesses and strengths of each plant. Sometimes a particularly lovely flower will catch my eye and distract me from the fact that the plant itself is gawky and ugly, or the fact that each individual flower is small will keep me from noticing how fragrant and profuse they are. Scoring each trait individually slows me down and helps me

Sharon Kardos of St. Clair Shores, Michigan, started breeding roses in 2005, inspired by online friends who sent her some seeds to start with, and the fun of that first experience hooked her on breeding.

You can't imagine how exciting that prospect was for me. I followed all the instructions and carefully cleaned and stratified the seeds. Every Sunday morning the ritual was, make coffee and go check the seeds to see if any had sprouted. At about the six-week mark I was starting to give up all hope when, lo and behold, I found one tiny white root sticking out of a seed. That was probably the biggest thrill of my rosy life! I had my first very own new rose. I tenderly planted her into the seed tray and off she went, putting on leaves. And off I was on a new, fun, and exciting hobby. I still have that first seedling. It took her three years to give me a bloom, but she's in the ground, six feet tall and four feet wide now. She looks a lot like her mom, but she's mine, and every June when she blooms she fills me with joy.

All gardening, and maybe breeding in particular, is about hope and looking to the future. Both things that are good for the soul. And there is something about growing things that just cries out to be a shared experience. Even though I know some of them only through the ether world, I have "met" some wonderful people I consider dear friends through gardening. That can only be a good thing.

really see, and down the line, allows me to easily plan new crosses between individuals with complementary strengths and weaknesses.

For example, last year I crossed one of my favorite modern roses, 'Strike It Rich', which has lovely deep golden yellow flowers, and one of my favorite older roses, 'Madame Ernest Calvat', which is a rather boring pink, prone to disease, and doesn't rebloom very well, but which I still love deeply because it is simply the most fragrant rose I've ever smelled in my life. My original goal was great fragrance with better rebloom, yellow color, and more disease resistance. The seedlings, thanks to the genetic heterozygosity of their parents, all look very different, but most of them have decent disease resistance and excellent fragrance. They're all pink, however. I got some of what I was looking for, but I was also very happily surprised to find one seedling that also boasts incredibly beautiful silvery-blue leaves, which is beyond exciting. Because roses, like virtually all perennial, clonally propagated crops, are heterozygous, the first-generation hybrids show a lot of diversity, but don't be fooled into stopping there. The F_1s, for all their diversity, still show only a fraction of the variability I'll see in the next generation. I've picked out my favorites of the F_1 plants, the one with the lovely foliage, and the most fragrant, deepest colored, best reblooming individual, and I'm crossing them together to get to the F_2 generation, where I'll hope to see some yellow blooms, better disease resistance and fragrance, and some more beautiful foliage.

Disease resistance and stress tolerance

With my rose breeding, I also manipulate the environment where they are grown, but, in contrast to what I want for my early tomatoes, my primary focus isn't to make sure their environment is perfectly uniform; rather, I try to make sure they get disease. This is a technique used very effectively by Bill Radler, creator (as an amateur in his backyard, by the way) of the deservedly popular 'Knock Out' roses. His technique was to collect the diseased leaves of roses, throw them in his blender with water, and then spray the mix over the leaves of his seedlings to make sure every plant got exposed to the disease. By doing this you get to see which plants are *really* disease-resistant. The results can be saddening at first, as previously flawless plants succumb to disease, but the final results—as you can see from the 'Knock Out' series—are well worth it. Be a little tough on your varieties. If you coddle them with the best soil and most careful watering, you risk breeding varieties that demand that treatment. If you normally irrigate, cut back in your beds with breeding populations and see which individuals take it best. Don't use any pesticides. Let them tough it out, and you'll breed hardier, easier-to-grow plants.

Flavor

Let's stroll away from my roses and back to the vegetable garden, to take a look at a project I am just getting started with. Radishes. I got interested in them when I first tasted the huge, mild Asian daikon radishes, so radically different from the tiny, round, pungent varieties I was familiar with. So I've decided to start exploring the group a little. As with all my vegetable breeding, exceptional flavor is one of my main goals, but how do I evaluate it? Most of the time, evaluating flavor is pretty simple. Pick a tomato from the plant, taste it, and if it is good, you take another tomato from the same plant, and save seeds from it. As long as they have been self-pollinated, the seeds from one fruit won't be different from another. Similarly, with corn, I can harvest one ear on a plant early to taste it as a sweet corn, then let a second ear mature for seed. Radishes, however, are a challenge, as are some other crops like onions and carrots. Each plant makes a single root, and to test the flavor, you have to uproot the plant and eat it. It doesn't do any good to note that a particular radish is exceptionally delicious if it's now in your stomach, busily getting digested. For carrots, I've found that I can cut a small sample from the very bottom, replant them, and still have them flower and set seeds for me as usual. I'm just now making my first crosses with radishes, and I'm not sure if I'll be able to approach them the same way.

For any plant where you can't figure out a way to evaluate it without killing it, you simply have to delay your selections to the next generation. Instead of evaluating and selecting your F_2 plants, save seeds from all your F_2 individuals separately, and plant them out in separate blocks the next year. You can then taste several individuals from each F_3 group, confident that the uneaten siblings of exceptionally delicious individuals will be pretty tasty as well. It is good that radishes are very small and quick to mature, because this process obviously can take up a lot of space. To limit the amount of garden real estate this project takes up, I'll be doing selection for as many traits as I can the first year—overall health, vigor, size, color—and limit the flavor evaluations in the next generation to the offspring of individuals with high marks on those traits.

Collaboration

As I walk through my garden, I also see a lot of collaborative breeding projects that I'm working on with other people. I've got snapdragons that I'm going to be sending to a friend in a warmer climate, so we can see how heat tolerant they are, and gorgeous kale from seeds another friend sent me. I'm just getting into breeding kale, while he's been working on it longer and very kindly shared some of the cool stuff he's created. I hope I'll be able to return the favor in a few years. A bed of dwarf bearded irises tucked away in a corner is the result of long conversations with yet another breeding friend. We got on the phone and chattered away about different directions a breeding program could take. I spouted off my ideas, wrote a check to his nursery asking for a few plants that we'd discussed . . . and a box arrived packed to the brim with far more than I'd paid for, oodles of fun to play with for a long time to come.

I feel that I've only scratched the surface of the potential for incredibly enjoyable, productive collaborative breeding projects. The rise of the Internet and gardening and breeding message boards have made it easier than ever to organize around the country and the world. The potential to connect with others—whether to informally exchange breeding material and ideas, or for more formal, organized group endeavors like the Dwarf Tomato Project—is huge, and the mechanics of doing so easier than ever. Get involved in a community garden or an active garden club; they too offer great opportunities for group breeding projects. Being able to adapt to your local environment is one of the best things about doing your own breeding, and working with people in your community makes it possible to leverage multiple minds and garden spaces to produce something truly perfect for your local gardens. What a project if everyone in a community garden grew a few tomato seedlings then all got together to taste and evaluate them and created something new, named for the garden itself! Or if a whole garden club got together to grow out and breed their own petunia. What could be better for a spring plant sale fundraiser than varieties bred by members of the club itself? The possibilities are endless. Get out there and start making them happen!

Naming and sharing

The final steps of selection are to name and share (preferably in that order) what you have created. Both steps are, I suppose, technically optional, but they are a great deal of fun and really add to the pleasure you get out of breeding plants.

Names are important and are forever getting confused—or causing confusion. Before you give a variety to another person, I strongly recommend you give it a name: gardeners love to share plants, and every time they do, that same plant, if unnamed by you, will take on some new moniker; a few years down the line, there will be a dozen people, all growing the same thing and calling it something different ("Joseph's tomato," "Tomato from Michigan"), leading to confusion and frustration all around. Come up with a new, distinctive name, and there is a better chance everyone involved will know what they are growing and not end up trading with someone else for a cool variety, only to find it is the same thing they already had.

When settling on a name, be sure it is unique. Particularly in the heirloom vegetable world, there are quite a few (far too many) different varieties all going about under the name of 'Brandywine'. A quick online search for the name you want to use is a good basic way to screen for synonymous (or nearly so) plants. Speaking of which, you'll also want to avoid giving your creation a name that is too similar to an existing variety. Name your new variety 'Bob's Brandywine' and you can bet somewhere down the line someone will get lazy writing the label and leave off the first half, and you'll end up contributing to the confusion of 'Brandywine' tomatoes.

To mitigate the chaos of overlapping names, some groups of plants—irises, roses, and daylilies, for example—have registering organizations that keep track of the names of all introductions. Generally, if there is a major national organization around a genus, they will have some sort of name registration system. The process of registering a variety will vary from group to group, but generally a quick Internet search will track down the information you need. There is no requirement that you register your variety names, but if you are going to share them with others, and certainly if you are going to sell them, it is a very good idea. Plant name registries make sure names don't get duplicated and provide a central place where you and others can get information about a particular variety. For vegetables and less popular groups of ornamentals that lack a registering organization, do what you can to make sure the names you use are unique and distinctive. Putting a brief description and some photos on a blog (or somewhere else online search engines can find it) is also a very good idea, so that anyone down the line who happens upon your plant can find out more about how it was created.

Once you have named your new variety, there are many ways to share it with other people. Obviously you can simply give it to friends and neighbors. You can offer your new selection through the Seed Savers Exchange (see "Plant Sources"); or you can trade with other passionate growers and breeders on one of the many online sites devoted to sharing information and varieties of various groups of plants (see "Useful Websites"). Whether you are sharing or trading, provide as much information as you can about where and how your variety was selected, so other gardeners will have a good idea how it might perform for them and why they would like to grow it. Try to be reasonable in your evaluation of its virtues, as well. Everyone tends to think their own babies are the smartest in the class, but that isn't always actually the case. Remember too that what you bred for yourself in your backyard will be adapted to your climate, soil, and tastes, and that may not translate to great performance in another person's garden, or a great flavor on another person's tongue. Before you boast that you are growing the most delicious tomato in the world, have a couple of friends do a blind taste test comparing it to other delicious varieties to make sure it isn't just you; be specific about its attributes ("mellow" or "tangy" both travel better, say more, than "delicious"). Providing a thorough, honest evaluation of your new variety, in whatever forum you offer it, will make sure it finds a place in the gardens of other people who will truly enjoy it.

Selling your creations

It is possible, at times, to sell the varieties you have created. As I have emphasized, breeding for the mass market and breeding for yourself are very different things, and many of the most wonderful varieties you create will simply not be marketable; but it is also true that the gamechanging 'Knock Out' roses, the beautiful hybrid hollies of *Ilex* ×*meserveae*, and many other very popular, successful plants were bred by amateur breeders. If you think you have something truly exceptional, it might be worth introducing and selling on the larger market. From the outset, however, face the reality that few plant introductions will make you much money. It can be incredibly satisfying to share the varieties you have created with other gardeners, but don't be under the illusion it is going to make you rich unless you are both very, very talented and very, very lucky.

If you really want to introduce a plant variety to commerce, it is no time to indulge in sentiment. Be ruthless. Get other people to help you be ruthless. Get online, search very carefully through available varieties, and compare them to what you've created. To be commercially viable, your variety must be a real, dramatic improvement over what is already in the market. Experienced gardeners probably just read that and laughed. After all, there are hundreds of nearly indistinguishable varieties on the market in popular, profitable groups like petunias or sweet potato vines. The reality is that the rules for big companies are different from what they are for you as a backyard breeder, and corporations often breed varieties very similar to what another company produced just to compete. For you, a small-time breeder, to hit the market, you need to think of something different, something so dramatic it is worth their time to invest in it.

If your variety is all that, you have two main courses of action. The first is to contact a plant breeders' agent. Plant breeders' agents work by organizing the trialing, patenting, and marketing of your plant; they are compensated by taking a percentage of the royalties earned from the sale of the patented variety. For this course of action to be practical, the variety has to be something truly amazing, capable of selling on massive scales around the country to make the expense of patenting and marketing worthwhile. The second option is to approach a company directly about selling your variety. Contact a smaller seed company or nursery about your plant and work out an agreement with them to sell your variety. Smaller, specialty nurseries that you love to buy from will probably be most interested in the interesting varieties you've created, but don't expect to make much money off the arrangement. These small companies operate on already tiny profit margins, but the pleasure of seeing the variety you created out in public—and earning back a little bit—may make it worth it.

*

In the end, no matter how or if you share your variety with other people, the process of evaluating and selecting a new favorite is incredibly pleasurable. There is nothing quite like a salad of fresh vegetables you bred yourself, or handing a loved one a full bouquet of a rose you've named for them. In this chapter, I've laid out some of the techniques I use, guidelines on approaches you can take with different types of plants. But remember, these are just guidelines. In the end, it boils down to getting out there, sizing up what you've got, and picking what you like best. I love dreaming up elaborate selection schemes, complete with charts and scores for various traits, but once I actually get into the garden and look at the plants, all that usually goes out the window, and I focus in on that plant or several plants that really catch my eye or nose or tongue, those special plants with that ineffable something that makes me love them. Trust your gut. As a gardener, you know what a happy, lovely, wonderful plant looks like. Find those plants, grow them in your garden, and share them with your friends.

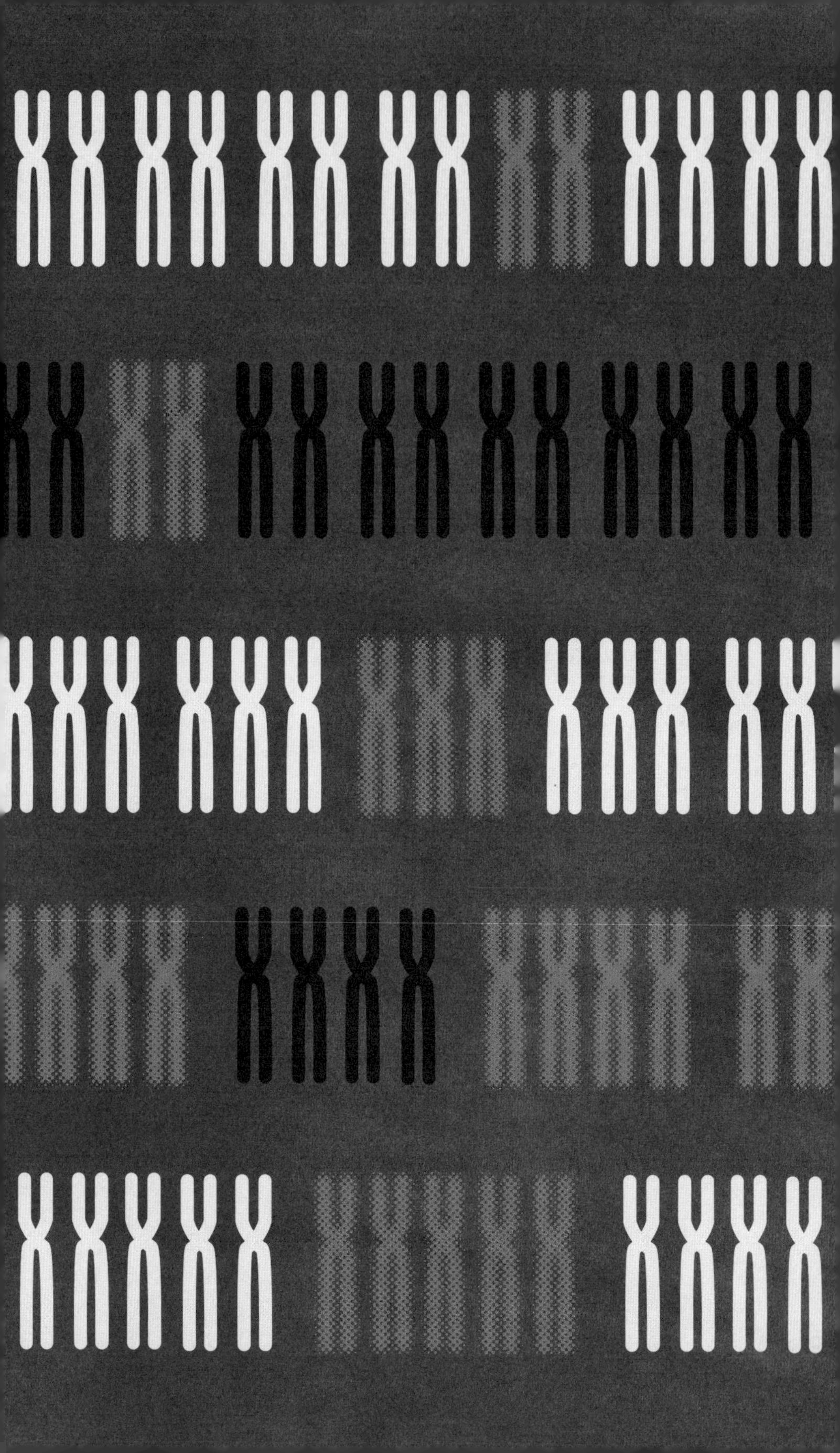

Beyond the Backyard

ADVANCED
TECHNIQUES

6

Threading the basic techniques of picking parents, moving pollen around, and evaluating the offspring is still the heart of all breeding efforts in the world. But at the highest levels—massive international companies breeding economic powerhouse crops like corn and soybean—and in specialized nooks and crannies elsewhere in the horticultural world, a range of additional tricks and technologies are being used to speed up and augment the traditional tools of breeding. Some of these techniques can be attempted, if you're adventurous, but most are impossible to use at home, requiring expensive equipment and training. What's more important is that understanding the principles behind how they work will give you an educated take on the politically charged issue of genetic engineering and an insight into the origins of some commercial varieties.

POLYPLOIDY

Polyploidy is the condition of having one or more extra sets of chromosomes. Most plants and animals, including us humans, are diploids, meaning we have two copies of each chromosome, one inherited from each of our parents. But many plants are tetraploids, meaning they have four copies of every chromosome, two from each parent. Tetraploidy can happen naturally, and many plants have both tetraploid and diploid versions in the wild; a familiar example of a natural diploid-tetraploid pair would be *Vinca minor* (diploid) and *V. major* (tetraploid).

Tetraploidy can also happen unnaturally. Scientists have learned to induce tetraploidy by using certain chemicals—most famously colchicine but also some herbicides like trifluralin. When applied to a growing bud, these can cause the chromosome number to double, producing a new tetraploid growth.

The line between creating a tetraploid and killing the plant is a fine one. Most of the chemicals used in the process are quite toxic. Though I do know some amateur breeders who have attempted to create their own tetraploids, most of you won't be trying it. But you almost certainly are already growing tetraploids that nature or science has created: many plants growing in your garden, from potatoes to roses, are tetraploids. So it is worth understanding a little about how they work. The diploid to tetraploid transition is useful for several reasons.

Bigger, better, stronger

Tetraploid versions of diploid plants are usually bigger in all their parts (thicker leaves, thicker petals, larger flowers—think of that pair of vincas), are often more vigorous, and show a wider range of variability than the diploids they came from—usually. Exceptions to this rule are numerous, but when tetraploidy works as planned, the effect can be quite striking and useful. Hostas, irises, and many other ornamental genera have been transformed this way; if you open a specialty daylily catalog, you'll see many, many tetraploid varieties. To some people, the larger, thicker petals of tetraploids lack the delicacy and grace of the diploids, however. In this, as in everything, each person's taste is different.

Restoring fertility

Many times, two different species can be hybridized, and the hybrid will grow perfectly well, apparently healthy and happy; but when it comes time to make the next generation, it's a no-go: they are completely sterile. The horse-donkey hybrid we know as a mule is the classic example of this. One of the main reasons for this sterility is that the chromosomes from the two parents are too different. The key step to making sex cells, whether they be sperm or pollen, is chromosome pairing. In your body, remember, you have two copies each of 23 chromosomes, one copy from your dad, and one from your mom. When you make eggs or sperm, the two copies of chromosome 1 recognize each other and pair up, the two copies of chromosome 2 recognize each other and pair up, and so on, up to chromosome 23. Once all the chromosomes have paired up with their partner, the 46 chromosomes divide in half, forming two new cells, each with just one copy of each chromosome. This recognition and pairing up before cell division ensures each sex cell gets exactly one of each chromosome, and so isn't missing some key genes because it didn't happen to get a copy of chromosome 6. In hybrids between two very different species, however, chromosomes of the two parents are sometimes so different they can't recognize their partners to pair up and divide evenly, and the whole process breaks down. The poor mule is an even more extreme case, as its two parents have different numbers of chromosomes, leaving it with 63. Try dividing that evenly by two and . . . well, the long and short of it is, no babies.

Making a tetraploid solves all this by simply doubling the number of chromosomes. Take the hybrid foxglove, *Digitalis* ×*mertonensis*. One parent is *D. grandiflora*, a beautiful perennial yellow species, and the other is *D. purpurea*, the classic purple/pink biennial foxglove. Their hybrid is sterile, until it is made into a tetraploid. Then, instead of having the lonely *D. purpurea* chromosomes looking fruitlessly for their mate among all the *D. grandiflora* chromosomes, suddenly you have an extra copy of each chromosome, so the *D. purpurea* chromosomes can pair up with *D. purpurea* and *D. grandiflora* with *D. grandiflora*. Everything comes out even, and perfect fertility is restored. One odd side effect of this, however, is that the new tetraploid, instead of producing a wildly diverse F_2 population, will now come true from seed. Because the two sets of chromosomes don't interact, their respective genes don't get mixed up into new combinations, so the next generation will look essentially the same as their parents, with exactly half their genes from *D. purpurea* and half from *D. grandiflora*. This is both a good and a bad thing. If you like the traits you get in the F_1, you have an instant variety that comes true from seed. If, on the other hand, you were hoping to dig into all the fun diversity of the F_2, you are out of luck.

It is worth noting that in the case of this digitalis, no one made this happen by treating a plant with chemicals; the switch to tetraploid was spontaneous. This does happen now and again, so if you have a sterile hybrid, and it ever does set seed, be sure to collect them. You just might have discovered the fertile tetraploid version of it.

Creating sterility

I know I just said tetraploids can be used to restore fertility, but they can also be used for the opposite effect and to make a plant sterile. This is how seedless watermelons are bred. All you have to do is take a tetraploid and breed it with a diploid. The hybrid will get two copies of each chromosome from the tetraploid parent, and one from the diploid resulting in . . . three. That's called a triploid, and with three copies of each chromosome, there is no easy way to divide things in half, resulting, usually, in sterility. In addition to seedless watermelons, sterile triploids are also being created to prevent potentially invasive ornamentals like buddleia and euonymus from producing seeds that can move into wild areas. The success of this has been mixed, as most triploids are not completely sterile, only very nearly so (you've found an occasional seed in a seedless watermelon, haven't you?), so often the "sterile" varieties can still send seed off to invade native areas.

Triploids are also used in ornamentals to prevent a plant from setting seed, resulting in profuse flowering without deadheading. The most successful use of this I know of are the triploid sweet alyssums, which are dramatically more floriferous than their regular diploid versions. (Though I still plant the diploids, because I like the way they self-seed.) Finally, triploids are a way for companies to protect their breeding. If all they release are triploids, it makes it that much harder for a competitor to use their varieties in breeding something better.

EMBRYO RESCUE

Embryo rescue is the final word in making difficult hybrids between distantly related species. Basically, many pollinations are initially successful; the seed begins to develop but then, before it reaches maturity, it aborts and falls off the plant. In embryo rescue, that partially developed seed is taken and put in a sterile gel containing all the food and nutrients needed for it to continue to develop and mature into a plant. The technique is the same as is used for tissue culture propagation of plants. Absolutely sterility is required to keep molds from killing everything off, and a very steady hand to handle the tiny, tiny, developing embryo in the failed seed, but with the proper equipment and *lots* of practice, you can do this at home. Embryo rescue is probably the most powerful method to break down barriers between species and get new hybrids. It is the technology that has produced all the exciting new intersectional lily hybrids, like Orienpets, which can't be hybridized any other way.

GENETIC ENGINEERING

Genetic engineering, the only advanced technique that sidesteps sex altogether, is by far the most radical and controversial of the beyond-the-backyard breeding methods. Traditional breeding introduces genetic diversity by bringing together all genes of two individuals, most of which have nothing to do with specific breeding aims. Genetic engineering instead inserts only one gene, one specific section of new DNA, into a single individual's genome, making it very precise. It also has essentially no boundaries.

While corn can breed naturally only with itself and a handful of closely related species, a genetic engineer can insert into corn genes from fish, bacteria, fungi, or viruses. That versatility makes the technology incredibly powerful. Increasingly genetic engineering dominates breeding of major field crops; it has been used to make plants that resist specific herbicides or are immune to specific viruses, plants that produce insecticides, or more Vitamin A, or flowers in new colors. Opponents argue that it is too powerful, and hence too dangerous, with great potential to seriously disrupt natural ecosystems. Proponents argue that we need to use it precisely because it is so powerful, that it could potentially help us develop more nutritious food to fight world hunger. Like any powerful technology, it needs to be handled with care, with a serious look at both the good and evil it could do.

MARKER-ASSISTED SELECTION

Marker-assisted selection is an increasingly important tool used in commercial breeding of major food crops. It doesn't really involve doing anything different; rather, it just does the same thing as traditional breeding, only radically faster. The short story is that our knowledge of genetics has increased far enough that for an increasing number of the big, economically important crops, it is possible to take a sample of a leaf, or even a chip of a seed, extract DNA from it, and then look at that DNA to see what genes it contains. Commercial corn breeders, for example, don't just plant out thousands of seeds to see what they will grow into. Instead, each seed gets sampled, analyzed, and a report goes back to the breeder so they can see what seed will grow into what. Instead of having to grow all the seeds, and select the individuals with the best growth habit and yield, they can plant only the subset of seeds they know have many of the genes they want, and select from them. They could get all the same information by growing each seed to maturity, but looking at the DNA ahead of time saves time and space.

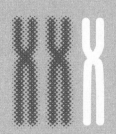

For Example

HOW TO BREED SELECT FLOWERS & VEGETABLES

7

Every plant is a little different, with different challenges, opportunities, and idiosyncrasies. In this chapter I'll go a little more deeply into the specifics of breeding some popular flowers and vegetables. Even if you aren't interested in breeding these specific plants, reading through some of them will give you a better idea of how the general principles I've explained in the previous chapters apply to particular situations. The final section of this chapter discusses how best to explore breeding with a new group of plants.

COLUMBINES

I'll always have a special place in my heart for columbines. One of the many factors that led me into gardening
were the columbines that popped up here and there
around my parent's porch in northeastern Ohio when I
was a child. They were typically pale pink, but one year,
I noticed one which was a much darker shade. I remember
wondering if it had crossed with the wild red-and-yellow
Aquilegia canadensis that grew along the stream down in
the woods and carefully collected seeds from it with
some idea of growing them, though I had only the vagu-
est idea of how to go about doing so at the time. I did suc-
ceed in germinating some and planting them in the new
garden I eventually created there, and they bloomed in
beautiful shades of pink and red, self-sowing happily, and
providing me with the chance to select out my favorite
colors. Since that day, I've always considered columbines
an absolute garden essential. After all, they are beautiful,
tough, and astonishingly diverse. Individual plants tend
to be short-lived, often little more than biennials, but they
self-sow so reliably you'll hardly notice.

Columbines are one of the very few groups of plants
with flowers in colors that pretty much cover the entire
spectrum—true blues, red, and yellow and every possible
shade between, including green, brown, and even purples
so dark they verge on black—and two or more of those
colors are often combined in the same flower for a lovely
multicolor effect. The flowers are complex: they can
have long graceful spurs or completely lack them (in the
clematiflora types); they can be elegantly single or show
masses of closely packed petals in their many double-
flowered forms. There are several species and hybrid

strains that even boast a strong, wonderful fragrance. In addition to their floral display, the leaves are delicately cut, and their standard green can be found streaked with white or yellow, or in a solid brilliant yellow-chartreuse that glows in a shady garden. I've grown plants that reach a graceful height of three feet in full bloom, and tiny dwarf forms of only a few inches. With all that diversity, there is a columbine for everyone, and your breeding goals need be limited only by your imagination.

Columbines are famously promiscuous. If you grow several different kinds in your garden, they *will* interbreed and self-sow their love children anywhere you let them. Because of that, if you grow columbines, you'll be breeding them whether you want to or not. If you leave them to their own devices, natural selection tends to drive them to the types most preferred by bees, short-spurred flowers in the purple to pink color range. If you want long graceful spurs and other colors, you need to get out there when each batch of seedlings starts flowering and be ruthless about ripping out the ones you don't prefer.

If you want to take more control, crossing two specific plants that you like, or want to maintain two or more colors without them interbreeding, you will need to hand pollinate. To be totally certain you are getting the cross you want, you should probably cover them to exclude bees, though I usually skip this because I find it hard to tie anything over the delicate stems without breaking them, and in my experience, if I remove the petals from the flower before it opens, bees aren't attracted to it and don't come by with any unwanted pollen. As always, the specific bees and conditions in your garden may be different. If you are making crosses, the large flowers are fairly easy to emasculate, though there are many anthers, and it takes some time to be sure you've gotten them all. Seed production is generous, and almost all pollinations take, so you don't have to do very many.

Columbines are natural outcrossers, as you can guess from their sheer promiscuity in the garden, so maintaining larger populations will help limit inbreeding depression. You can also propagate exceptional individuals by dividing them in the spring, though your success in maintaining a particular plant will depend on how much it wants to perennialize. Some forms are little more than biennials, while others, with regular division, can persist more or less indefinitely. Breeding for reliably perennial varieties is a worthy goal but also a fundamentally difficult and time-consuming one, simply because there is no way to evaluate how long a plant will live without waiting several years to see which individuals persist. To help me select for longer-lived plants in my garden, I recently started labeling each seedling with a color-coded tag, so I know what year they germinated. That way, over time, I'll be able to see which plants persist the longest and use them more in my breeding.

Columbine seeds store well and germinate easily with no special treatment, though most species and varieties will germinate faster and more uniformly if you give them a couple weeks of cold conditions first. For crosses I really care about, I put the seeds (and a moist coffee filter) into a plastic bag and refrigerate them for about two weeks before planting them as usual. Since many seeds will germinate without a cold period, however, I often skip the cold step if I have a lot of seeds and sow them directly. I figure by doing so, I'm slowly selecting for seeds that germinate well without any cold period, which will make my life easier in the future.

When it comes to making breeding goals, the sky

is the limit. Commercial breeding in recent years has mostly focused on extremely dwarf plants with upward-facing flowers (ugh . . .), but there is *so* much else to play with in this genus! Columbines have such a great color range, I'm always surprised no one has pushed it even further. It is surprisingly hard to find good varieties in the yellow-orange-red range, or even a mix of ravishing shades of blue. There is absolutely no reason we couldn't have columbines in all sorts of dramatic, elusive shades like the palest possible peach or dusky gray or both at once in the same flower. If you can imagine it, columbines can probably do it.

Fragrance is another great frontier. For years, fragrant columbine flowers were limited to a handful of finicky species, but thanks mostly to some brilliant breeding by Plant World Seeds in the UK, there are now wonderful, easy-to-grow strains with large, powerfully honeysuckle-scented flowers. They're still hard to locate in the United States, but several British seed houses supply them and ship internationally, so wherever you live you can grow these wonderful varieties. That fragrance is just dying to be bred into other shapes, colors, and forms.

Foliage effects are also a potential place for future work. Variegated and yellow-leaved forms are available, mostly with fairly unexciting flowers in a mix of colors. Great colored leaves with complementary flowers would be amazing. Some species, notably *Aquilegia chaplinei*, also have exceptionally beautiful, very finely cut, ferny foliage.

Personally, in a fit of pique over all the lumpen little dwarf varieties, I've become obsessed with breeding giant columbines. I want them huge—tall enough to look *up* into the flowers. This may be impossible, but I'm going to give it a shot, and I'm sure I'll find all sorts of fascinating things I'm not aiming for along the way.

So start playing with columbines. They offer diversity enough to inspire the most advanced breeder, and for the rank beginner, they are a great place to start because, really, you don't need to do anything more than buy a few packets of interesting seeds, plant them together, and start picking out your favorite forms as they seed around.

COLEUS

This is a group of plants that keeps reinventing itself. Beloved by the Victorians primarily as houseplants, in recent decades coleus have enjoyed a huge revival with a deluge of new breeding and forms, in brighter colors and novel leaf shapes. The trend has been away from the older shade or even houseplant forms, toward robust plants that can grow in full sun but do not perform well in lower light conditions.

With any species, like coleus, with such a huge range of forms and cultivars, the first step in starting a breeding program is to get familiar with what is available so you don't end up recreating what already exists. The widest range of leaf colors is found in standard-shaped leaves and regular upright-growing plants, while the varieties with a trailing growth habit and more unusual leaf shapes have a more limited color range. Expanding the color range for those more unusual forms is a natural place to start. Another interesting possibility for these plants would be to breed for something beyond the full-sun bedding-plant usage that is now the standard. If you have a shady garden or like houseplants, better varieties for low-light situations would be an excellent goal. The flowers of coleus are another neglected area. The blooms are small and generally regarded as a bad thing, detracting from the beauty of the foliage. I mostly agree, but I have seen some varieties that have fairly attractive flowers. Selecting specifically for larger, showier flowers would be a dramatic reimagining of the ornamental potential of these plants.

Coleus are outcrossers, but since they are incredibly easy to propagate by cuttings, you don't have to worry about inbreeding depression. Though not tolerant of frost, they are easily overwintered on a sunny windowsill; of course, the varieties that tolerate lower light will look better during their winter stay indoors than the full-sun lovers.

All cultivated coleus are derived from just one species, what used to be *Coleus blumei* but has now been shunted over to a new genus, *Plectranthus scutellarioides* (after a pit stop as *Solenostemon scutellarioides*). These name changes are frustrating, but they reflect an increasingly exact understanding of how different species are related to each other on a genetic and evolutionary level. As annoying as I find name changes to be, they have a good side, as better classifications give us breeders better clues as to which plants might successfully hybridize with each other. If you want to be adventurous, it might be worth trying crossing coleus with some of the many beautiful *Plectranthus* species. If you succeeded, the results could usher in a whole new era of coleus wonderfulness.

Coleus flowers are tiny, and though it is possible to emasculate them, it takes a fine pair of tweezers, a very steady hand, and a lot of patience. Particularly the patience: since each successfully pollinated flower yields only a few seeds, you'll have to do more than just a couple of flowers. It is more practical to simply cover the flowers to exclude the bees, then pollinate the open blooms without emasculation. You'll get some selfing, but most of the seeds will be crosses, and it will save you a lot of time.

You can get even lazier than that, if you want. I talked to one coleus breeder with several very successful varieties on the market who told me that he doesn't bother to make controlled crosses at all. He simply grows a lot of varieties together, lets the bees mix them up, harvests the seeds, and sows it all together. The results are unpredictable, but the payoff is having lots and lots of seedlings to select from for very little work. If you go that route, you can also try to guide the crossing process somewhat by growing the plants you want to cross next to each other and well separated from other flowering coleus varieties. Bees will still carry pollen far and wide, but the majority of it will come from nearby plants.

Which crossing technique you use will depend on your goals and breeding approach. If you are trying to combine two specific traits—say, the color of one variety with the leaf shape of the other—you'll want to make controlled crosses. For more incremental improvement in traits like shade tolerance, massing together your most shade-tolerant varieties and letting the bees mix them randomly will give you excellent diversity to select from.

When choosing parents for making crosses, remember that in most cases, flowering in coleus is not a good thing, so though it may be tempting to use prolifically blooming varieties as parents because they produce a lot of seed, if you do, you'll end up with less desirable plants. This is particularly important to take into consideration if you are simply letting the bees pollinate at random. All else being equal, the most prolific bloomers will produce the most seeds and pollen, so their heavily flowering genes will be disproportionately represented in future generations. You can avoid this by harvesting seeds from each plant separately, and sowing the same amount of seeds from each, so that heavy bloomers don't take over your population.

Coleus is one of the most mutable, interesting species, and given the stunning range of colors and forms
already bred from it, I'm fascinated to see what creative,
innovative breeding might produce from it in the future.

DAFFODILS

There's a lot to love about daffodils, the classic flowers of
spring, and with an abundance of diversity, they have also
attracted a robust amateur breeding community. Daffodils, like dahlias, roses, irises, and other well-established
groups of ornamental plants, have a whole obsessive sub-
culture built up around them, with many serious daffodil
lovers focused almost exclusively on growing and breeding varieties that will perform well on the daffodil show
bench. This means an overwhelming focus on absolutely
perfect form and color of the individual flower, but less
emphasis on the vigor and presentation of the plant in the
garden. The vast majority of daffodil bulbs sold, however,
come through the dramatically different pipeline of the
mostly Dutch commercial growers, who put a premium
on varieties that increase rapidly to marketable size in the
growing fields but care far less (than the daffodil show
crowd do, anyway) about details of form and coloration.

What you love best will vary. Some of my daffodils
I like to grow as cut flowers for the house, in which case I
value more of the things show people value (though I'm
really not enough of a daffodil connoisseur to appreciate the intricacies of what makes for a truly great show
flower). But most of my daffodils I enjoy in the landscape,
where I want a nice balance of vigor, beautiful color and
form, fragrance, and prolific flowers. Like most things, the
best way to get what you love best is to breed it yourself.

Daffodils are divided into several different groups, which are primarily designed for and used by the people at the show bench. Frankly, as a gardener, I could care less if a variety is classified as a trumpet, or a small cup, or a large cup, distinctions determined by the relative length of the cup to the petals; some of the other divisions, however, have more dramatically different looks and are useful in finding plants that you like the best.

The modern daffodil is the result of hybridization between many different species, and in most cases, the different species and divisions of daffodil can interbreed, though some crosses will be mostly sterile. The triandrus and jonquilla groups, in particular, tend to be less fertile in crosses with other divisions, though exceptions to this rule abound.

Luckily, you don't have to wonder if most daffodil varieties are fertile or not, because the daffodil world has one amazing resource that other genera lack, in the form of daffseek.com, a website sponsored by the American Daffodil Society. You simply enter the name of any daffodil variety, and get back how fertile or sterile it is, see its ancestry, and a list of its descendants. (Other plant societies, please take note! We would all love more resources like this!) This amazing website can save you a lot of time and let you make smart decisions with your breeding. If you are planning a cross, you can see if someone else has already tried it, and if so, what results they got. If you fall in love with a flower, you can find out if it has been found to be sterile by other breeders in the past. Looking at the offspring of a variety gives you an idea of the sorts of babies it might have. Don't, however, let this tie your hands. Just because someone has made a cross before doesn't mean you might not get something different or better out of it—after all, faced with the very same population, two different people would probably select different individuals out of it. And remember, sterility is rarely absolute. Spend most of your time with plants that are fertile, but don't be afraid to make a few crosses that probably won't work. Splashing some pollen around doesn't take long, and the results can be exciting.

Daffodil flowers are large, and straightforward to emasculate and pollinate. Most, but not all, breeders emasculate their flowers before they pollinate; but if you don't have a lot of bee activity in the early spring, you can probably get away with skipping it. I do emasculate but don't cover my blooms. Your mileage may vary, depending on the pollinators in your area and how obsessive you are. Most breeders report that crosses take best when made on warm days. As with all plants, fresh pollen is best, but daffodil pollen responds quite well to storage. Kept dry at room temperature, it can stay viable for a couple of weeks; in the refrigerator, it remains viable for several months, meaning it is very practical to save pollen to make crosses between early- and late-blooming varieties.

You can also take the lazy approach and simply collect open-pollinated seeds produced by visiting bees. This is actually the approach used by many Dutch breeders. The results will be unpredictable, but if you have a lot of space to grow out seedlings, it can be an easy way to get some fun stuff.

Fresh seed, harvested when the pods begin to turn brown, or can be heard rattling inside the pod, should be sown immediately for best germination. Keep the young seedlings well fertilized and watered to speed up their growth, and you can see their first flowers in three or four years. Breeding daffodils is not a project for the impatient, but if you start making crosses every year, pretty soon you'll have new blooming seedlings to enjoy and select from each spring, and since daffodils are wonderfully small, you can pack a lot of seedlings into a relatively small area.

Since daffodils are easily, indeed almost effortlessly, asexually propagated, once you find an individual you like, you can just let it multiply, give it a name, and share it with your friends. It is best to register your new variety with the American Daffodil Society, for the usual reason of preventing the confusion of overlapping names, and so it can be included in the daffseek.com database, providing more information for future breeders.

Breeding daffodils, those classic heralds of spring, is a great way to get your gardening year off with a bang. The first daffodil blooms are already cheered wildly as proof winter is finally over; imagine how much more exciting it will be to cheer on the very first flower bud of a new variety you created yourself. Take a few moments, make a cross, and find out for yourself what that feels like.

DIANTHUS

This is one of the great neglected genera. There used to be literally hundreds of cultivars of carnations and pinks for the garden, with a whole vibrant community of highly competitive shows, and now you'll be lucky to track down a dozen varieties. The heirloom movement is starting to bring some of them back, and it is high time we gardeners got busy exploring the diversity of this genus. Incredible fragrance, beautiful intense colors, profuse flowers over a long season, ranging from tall forms perfect for cut flowers to alpine dwarfs that make tiny mounds only a few inches high. Many also have uniquely frilled flowers—these range from small teeth at the petal edges in most forms, to long, trailing threads in the most extreme version, the so-called bearded dianthus from Japan. In all their various forms, these are amazing plants for the garden.

As a group, dianthus tend to prefer excellent drainage, and many melt in hot, humid summers; but in such a diverse genus, this is far from universally true. I have many species that thrive in my heavy, wet clay soil, and nurseries in the South (see, particularly, Plant Delights Nursery) are offering an increasing lineup of lovely varieties that thrive in sweltering conditions.

This is a genus of outcrossers, but plants are almost always easily propagated by cuttings or division, with the exception of the short-lived biennial species. Making crosses is fairly easy; the only tricky bit to crossing dianthus is dealing with extremely double forms. The extra petals are derived from the stamens, so fully double types, like your classic florist's carnations, often have no pollen at all. There are a couple of ways around this: you can use them as the female flower in crosses with other, male fertile varieties, or you can look for secondary buds. In carnations, there is one large bud on the end of the stem, but smaller, secondary buds sometimes form off the side of that stem. Typically these would be pinched off to maximize the central bud, but leave them on: they often have fewer petals and some pollen. Some breeders pinch out the central bud as soon as they see it to maximize the number of side buds they have to try to get pollen from.

With single or semidouble varieties, you'll need to emasculate to ensure you don't get self-pollination. This is easiest to do if you first pull off the calyx—the green tube forming the base of the flower. This will allow the petals of even the most double flower to flop open and expose the actual business parts of the flower.

Once pollinated, dianthus set seed generously, and seeds germinate easily without any special treatment.

In a genus this large, there are several different species and hybrid groups to consider, each with different strengths, weaknesses, and interesting options for further breeding.

Carnations

Carnations derive from *Dianthus caryophyllus* and some hybrids with related species; you probably know them as those tightly formal, scentless, incredibly long-lasting, long-stemmed flowers at the florists. The group, however, has a lot more diversity than that, including many forms bred for garden display rather than the cut flower industry. Historically, carnations were known for their strong, spicy fragrance, but this has been bred out of most modern carnations in the quest for long vase life.

Pinks

This is an inexact term. "Pink" is often used as a generic term for the entire genus, but it most often refers to the old-fashioned garden pinks derived primarily from *Dianthus plumarius*. These are great garden plants, and I'm not quite sure why they fell out of fashion. Intense fragrance, great vigor, and the most intricately patterned flowers I know. The colors are limited to the white-pink-red range, but with flowers blotched and edged with gorgeous contrasting tones, you'll hardly miss the other colors. Pinks are confused in the trade with the old-fashioned garden carnations. The two groups are further confused because they have been hybridized in the past, though they cross with difficulty.

Chinensis

These are the little, short varieties you'll see sold sometimes alongside pansies. I'm not a huge fan: they are short-lived and intolerant of heat, and tend to be annuals (though I've had them overwinter now and again). I appreciate their large, intricately patterned flowers, but really, there are better garden plants in the pinks. The biggest virtue of these varieties is from a commercial production standpoint. They're compact, ship well, and bloom quickly from seed. The only reason I have these in my garden is because I've read that they can be crossed fairly easily with carnations and the one (very pale) yellow species, *Dianthus knappii*. There may be breeding potential in those large, intricately colored flowers, but they need a lot of work.

Sweet Williams

Such a classic, old-fashioned plant, *Dianthus barbatus* is typically a biennial, forming a rosette of leaves the first year, then stretching the second to produce tightly packed heads of small flowers in a range of colors and bicolors. Commercial breeding has pushed this to shorter annual forms, which, personally, I don't like at all; but there are also reliably perennial varieties in the trade if you look. This may be the most widely adapted species, thriving in places as diverse as Michigan, North Carolina, and Southern California. I think there is great potential for this species to come out of the shadows of history, particularly with the reliably perennial forms. There are references to it being easily crossed with several other species, including *D. deltoides* (maiden pinks), *D. superbus* (with incredible lacy, fragrant flowers), and the rock garden classic *D. alpinus*.

Other species

The genus *Dianthus* is huge, almost 300 species by most counts, and more and more of them are popping up in catalogs. You may know *D. gratianopolitanus* from the lovely and deservedly popular variety 'Firewitch' (this species, by the bye, is reported to cross with the classic old-fashioned *D. plumarius* pinks); I've been seeing the lovely frilly blooms of *D. arenarius* (which can cross with carnations) in more catalogs; and there are a host of tiny, mat-forming alpine varieties beloved of rock gardeners. Though I've tried to pass along what little information there is about what species will cross with what, the reality is, most crosses simply haven't been tried, so feel free to experiment and see what you can create.

This old-fashioned group of plants is long overdue for a revival and reinvention in new, wonderful, fragrant forms. In the huge diversity, there is something for every taste and climate. It is just up to you to start exploring and putting it all together into varieties that transform this genus from old-fashioned to trendy.

HOLLYHOCKS

I love hollyhocks—such a classic, cottage garden plant, such gorgeous flowers. I know so many people who love hollyhocks, and yet so few who actually grow them. Hollyhocks, like so many tall, graceful plants, have fallen victim to the vagaries of modern plant marketing. Out of bloom (the stage at which it should be sold) it is an unexciting-looking mound of leaves, and in bloom, it is impossible to ship, a towering spire that is striking in the garden but looks awkward and ungainly in a pot on a bench. So hollyhocks, once so beloved, are an increasingly rare sight in gardens.

Another damning feature of old-fashioned hollyhocks is their biennial growth habit. Being biennials means the first year they're simply a mound of leaves, then the second year they shoot for the sky, produce stunning stalks of blooms, and die. Somewhere between marketing ever-blooming annuals and plant-and-forget perennials, biennials have been forgotten, but they still have much to offer. Because they spend an entire year building up energy then throw everything they have into flowering, saving nothing to survive to the next, biennial hollyhocks (and other lovely but increasingly forgotten species like old-fashioned foxgloves and chimney bellflowers) put on a show like nobody's business, with the wild abandon of blowing it all in one great display. It also isn't really hard to have a display every year. Hollyhocks love to self-seed, so as long as you don't go too insane with the mulch, they will happily settle in, seed around, and bloom annually with little work from you. If you want to get a more controlled display, you can sow them in rows in your vegetable garden in the summer, transplanting them in the fall or spring to where you want them to bloom. It is a little work, but the sheer impact of the flower effect can be worth it. Besides, no-work gardening is overrated. There is something immensely satisfying about doing a little planning and spending an hour with a shovel to engineer something truly stunning.

If biennials aren't your thing, there are other options. Newer seed strains of hollyhocks will bloom their first year from seed as annuals, and other species and hybrids are soundly perennials. But I'd encourage you to try the old-fashioned biennials. They'll never be commercially mainstream because they don't fit neatly into the production modes that have been developed for annuals and perennials, but in the garden, biennials are easy to handle and wonderfully beautiful.

The other problem with hollyhocks is the fungal disease rust. Rust will rarely kill a plant, but it will disfigure the leaves, particularly those at the bottom of the plant. It is less of a problem in drier climates, and planting hollyhocks where they get good air circulation to keep the leaves dry will help. But in my wet climate (and given my habit of packing as many plants as possible into a bed) it is a something of a plague. I more or less accept it at this point. As tall as they are, hollyhocks go in the very back of my bed anyway, so I've always got lots of other things in front of them to cover up the rusty foliage. The classic hollyhocks, *Alcea rosea*, are confirmed rusters; several other species, notably *A. rugosa* and *A. ficifolia*, are somewhat more rust-resistant, but nothing I've grown is actually immune. Your experience will vary, according to your climate.

Classic old-fashioned hollyhocks come in a range of pinks, reds, pale yellow, and warm, salmony melon shades, plus white and the ever-popular dark, dark reds that get sold as black. These are most commonly available only as a mix of colors, which is not very useful if you like to carefully design your garden color scheme. Selecting out superlative single-color strains in your very favorite hues would be a quick, easy, and very rewarding breeding project.

Hollyhock flowers come in either the classic single, or full, ruffly doubles, the relative virtues of which gardeners can debate forever. I like them both (don't tell anyone . . .)—I enjoy the lush, ruffled fullness of the doubles and the simplicity of the singles equally—but I know many people who loathe one or the other. I also enjoy some of the single-flowered strains that boast lovely bicolored flowers, with contrasting darker or lighter flower centers or edges. There is much to be explored here, in pushing more exciting, complex color forms.

The overall look of a hollyhock plant is another area with lots of room for improvement. In old, self-sown stands of hollyhocks, I frequently see some plants with prolific flowers blooming up and down the entire spike at the same time, while others look spotty and moth-eaten, with flowers scattered here and there. Personally, I detest staking, so though I like my hollyhocks tall and dramatic, I also like them with very strong stems that can stand up on their own. If you don't mind staking, however, with a little support, you could breed them to reach truly spectacular heights, something sadly lacking in many modern garden perennials. There are some commercial dwarf strains, clocking in at less than two feet, but I see utterly no point to them and am sure you'll avoid the loathsome things.

A few species of hollyhocks are widely available, most commonly as seeds, which are large and germinate into vigorous, healthy seedlings quickly and easily.

Alcea rosea is the poster child of the genus. Huge spikes of flowers in almost any color except blues, including a near black. Single flowers, double flowers. Mostly biennial or short-lived perennials, they love to self-sow if you let them. Prone to rust. There are a handful of newer strains, but most of what you'll find on the market are older varieties in mixed colors.

Alcea ficifolia and *A. rugosa* are both more perennial and more disease-resistant than *A. rosea*, though neither is perfect in either regard. They've both been hybridized with the old-fashioned *A. rosea* to lovely effect, producing vigorous plants that are, to my mind, far superior to the originals, though the range of colors and forms available in the hybrids is still somewhat limited; for example, I don't know of any double-flowered hybrid types, so there is still a lot of interesting work to be done crossing these healthier hybrids with the most beautiful of the *A. rosea* varieties.

In addition to these three more common species, some of the approximately 60 other species of hollyhocks are occasionally available from specialty seed suppliers. There is at least once incidence in which *Alcea rosea* has been successfully crossed with the closely related marsh mallow, *Althaea officinalis*. The resulting hybrid is beautiful, apparently disease-proof—and unfortunately sterile, but the potential is there to make some really lovely things by recreating the cross with other interesting forms of *Alcea* and *Althaea*. With so much unexplored diversity, there are surely any number of gorgeous plants yet to be discovered.

Hollyhocks are satisfying to work with as the flowers are large and straightforward to emasculate and pollinate, and each flower produces many large, easy-to-germinate seeds. Also pleasantly, most species of *Alcea* can cross, and the resulting hybrids are fertile. In my garden at least, bees show a great fondness for the flowers, even after I've pulled off the petals and emasculated them; so to be sure you are getting the cross you want, you'd be wise to cover the flowers before and after pollination.

Hollyhocks are a plant that needs some love from home breeders. Neglected and unsuited to mainstream commercial production, they are a perfect place for people like you and me to get busy creating something beautiful to share with friends and family.

ROSES

One of the oldest of cultivated flowers, roses have been grown in gardens both in Europe and Asia since the Middle Ages. Entire libraries of books have been written about the history and development of roses, and true enthusiasts will probably be horrified by the brevity of my treatment here, but this is their history, in the western world, in a nutshell.

The first varieties, derived from native European species, comprise most of the classic "old roses" with big flowers, densely packed with fragile petals, and strong fragrance. They bloom only once, at the beginning of the summer, and for the most part are very winter hardy and resistant to disease; their colors are limited to white, pink, and a vivid magenta-red. That all changed with the introduction of Asian species, particularly China roses, which bloom repeatedly all summer in shades of coral and pale yellow, in addition to pink and crimson. The color range of roses then really expanded with the intro-duction of blood from one of the wild roses of the Middle East, *Rosa foetida*, which finally brought true bright yel-lows to the world of roses.

Along with reblooming and bright, varied colors came less winter hardiness and the modern plagues of black spot, rust, and mildew. Interestingly, these maladies were not much of a problem at first: when Europe was still running on coal, the air was filled with pollution that (in addition to all the damage it did to the environment) acted as a fungicide and prevented the development of fungal diseases on roses. As the air began to clear, disease became a serious problem, and rose growers turned to an extensive array of fungicidal sprays to keep their plants happy. Today we've inherited the results of decades and decades of breeding for plants with enormous, gorgeous flowers in a huge range of colors that bloom all summer—but demand fungicides to stay healthy.

The long, dark years of spotty, wretched roses did their damage, however, and many gardeners who have had bad rose experiences swear never to plant them in their garden again. But the tide is turning, and a whole revolution of new roses is upon us. Everyone knows the disease-proof 'Knock Out' varieties, of course, but even in the typically prissy groups of Hybrid Teas and Floribundas, many newer releases grow fairly happily in my garden, where I never spray anything.

For many breeders, for many years, the focus had been almost exclusively on the beauty of individual flowers. Hybrid Tea and Floribunda breeding basically ignored the look of the plants as a whole and simply focused on the most stunning flowers possible. These plants tend to be gawky and awkward in a landscape but make great cut flowers. In fact, I grow them in rows with my vegetables and enjoy them exclusively in a vase. The modern focus, on the tails of the incredible commercial success of the 'Knock Out' series, is on disease resistance and vigor above all, on bushier, more attractive plants. Though I admire 'Knock Out' roses as a triumph of breeding I'm not a big fan of the 'Knock Out' flowers, and many of the new rose varieties I've seen being promoted by the big marketers I feel miss the point of growing roses. I want them to be disease-resistant, of course, but to my mind, if the flowers aren't big, voluptuous, fragrant, and decadent, what is the point of growing a rose? If I wanted dinky little scentless flowers on a tough plant, I'd grow potentilla.

You'll find that within such a diverse group as roses, there is an endless array of fascinating breeding projects to attempt and something to please (almost) everyone. Roses tend to inspire strong opinions. Some people love the rigorously formal flower of the classic high-centered Hybrid Tea, others the loose informal bloom of an old rose. I personally go weak at the knees for the cupped flower form seen in many of the David Austin varieties. Fragrance is also very personal. The classic Hybrid Tea 'Fragrant Cloud' apparently lives up to its name to most people, while my nose barely detects a scent in it. The myrrh scent of some old roses and many of the David Austin English roses smells lovely to some people, while it reminds others of urine. I personally think many Hybrid Tea roses smell unpleasantly of baby powder. A friend sniffed and agreed, but surprised me by saying that many people enjoy the smell of baby powder. So don't trust anyone else when it comes to the best-smelling roses. Spend time in public gardens and nurseries sniffing away, collect what you like best, and breed them.

Their floral beauty often blinds us to the fact that roses can also have foliage that is desirable for one reason (or season) or another. Many modern roses have new leaves that emerge a rich purple-red in spring before fading to green, setting off the flowers to stunning effect. Turning to some species lesser used in modern breeding, *Rosa glauca* famously has purple foliage overlayered with a thick coat of silver; *R. rubiginosa* and *R. pulverulenta* have strongly fragrant leaves scented of green apples and pine, respectively, perfuming the whole garden on a still summer's morning; and the leaves of *R. xanthina* f. *hugonis* are tiny, creating the effect of delicate lace, perfect to set off its delicate, pale yellow flowers that open weeks before any other rose. *Rosa nitida* has beautifully shiny leaves that look like they've been covered with a thick layer of lacquer, seen to best effect in its striking hybrids with *R. rugosa*. And the leaves of many species roses, including *R. nitida* and *R. rugosa*, turn vivid shades of red, orange, and yellow in the fall, bright enough to compete with the very showiest maples. Though long neglected, breeding for attractive foliage is a wonderful next-step in creating roses that are truly worthy of inclusion in every landscape.

Spend much time talking with rose people, and you will quickly realize that location is everything. Certainly this is true of disease resistance. Growers on the West Coast constantly battle rust, a disease I've never seen on my roses. In my zone 5 Michigan garden, black spot is a constant plague, but mildew is a rarity I hardly think about. In drier climes, the situation is just the reverse. 'Knock Out', a compact and attractive three- to four-foot shrub here, expands to a house-eating seven or eight feet in the heat of the South. English roses rebloom nicely in cool areas but tend to run to leaf in warmer climes like California. Varieties that slowly open gloriously delicate flowers in my climate fade and burn in Texas, while the Noisettes, which are robust essentials for the Deep South, sit and pout through my cool summers and demand careful swaddling to survive the winter. This is true of many plants, but for roses particularly, the best one for your garden is going to be one that was bred in your area. So by simply selecting for healthy, vigorous plants that grow well for you, you'll be able to create wonderful varieties that can outcompete all but the best breeding from other places.

Hybridizing roses isn't too complex. Most varieties will self-pollinate, so they do need to be emasculated, but with such large flowers, that is easy to do. In most climates, covering the pollinated flower will help keep the stigmas from drying out and increase your seed set. Personally, I never cover my roses, but in my cool, humid climate it doesn't seem to be necessary, and I've never seen bees or other pollinators visit a rose once I've removed the petals.

Roses can be a finicky bunch. Because so many species have gone into their makeup, they differ radically in their fertility. Some varieties are nearly sterile, while others set seed with almost anything you throw on them. HelpMeFind.com is an invaluable online database that shows the offspring and ancestry of almost all varieties, which can give you a clue to fertility; but talking with other breeders—and experience—will help you find the best parents for your climate. Though most species of roses will, with varying degrees of difficulty, cross, there is one big barrier to breeding roses: chromosome number. Most modern roses are tetraploid, while many wild species and Rugosa, Noisette, and China roses are diploid. Unlike many plants, the triploid hybrids between the two groups are often not completely sterile and so can be used in breeding. Some rose breeders simply ignore the issue, but know that if you cross between the two groups, you might be setting yourself up for disappointment. In addition to all its other excellent information, HelpMeFind.com lists chromosome numbers for most species and varieties.

Most roses take a very long time to ripen their seeds, so in cooler climates, you need to pollinate the first flush of flowers when they start in early summer in order to have them mature before fall. There are exceptions: some species of roses, particularly the Rugosa types, ripen their hips very quickly. Most roses won't successfully pollinate in extremely hot weather (100 plus), so in very warm climates, you'll have to schedule your pollinations in the cool of spring or early summer. Usually, I harvest hips once they begin to change from green to red/orange, though often seeds from a still-green hip will be viable. In my climate I leave things on the plant as long as I can and harvest everything, whatever its color, when my first frost threatens. If you have a very long season, you can confidently harvest after 12 weeks.

Rose seeds require a cold treatment of about three months to germinate, and you can expect germination rates to be just as variable as seed set. Some apparently good-looking seeds will never germinate, and others may take two or more winters to sprout. Again, numbers are your friend. Be prepared to make lots of crosses; not all the seeds will germinate.

Seedlings of modern roses will bloom quite quickly, often just a few months after germination. The first flower will be much smaller than those that will bloom on mature plants, and often has fewer petals as well. Color and fragrance are more consistent, but these too can change as plants mature. If space is an issue, you can select seedlings at the time of their first bloom, but if you have the room, growing them out to a mature plant will give you a much better idea of what they actually look like.

Once-blooming old roses and species roses can take much longer—several years—before they bloom. Repeat blooming often acts as a recessive gene, though in such a complex group of species there are many exceptions to that rule, so generally crosses between a repeat-blooming rose and a once-blooming rose will not repeat bloom and take several years to reach flowering size. This is frustrating and time-consuming and makes many breeders shy away from working with species roses, but the payoffs in terms of novel forms and increased disease resistance are great.

Modern roses are genetically diverse and heterozygous, so even the first generation of crosses will show a huge range of diversity. Remember, however, that crossing between those seedlings to generate an F_2 generation will reveal more diversity. This is time-consuming, of course, but may be useful if there are specific traits you want to recover from a cross, like rebloom or a certain flower color, that don't show up in the first generation.

I made my very first intentional hybrid with a rose, and I still have a soft spot in my heart for them. They're a thorny, sometimes needy group of plants, but they are also astonishingly beautiful, and there is a whole host of even more beautiful roses just waiting to be created.

SNAPDRAGONS

The common snapdragon is a wonderful old-fashioned plant, historically grown for its towering spikes of flowers in a lush range of colors and whimsically shaped blooms. Modern breeding, has, as with everything else, been focused primarily on dwarf forms, though taller varieties continue to be bred for the cut flower industry. In my climate, snaps happily bloom through the summer; in warmer climates, they are best grown as a winter annual.

In addition to the wide color range, several different flower forms beyond the classic snap form occur, including doubles and open-faced nonsnapping forms that I see absolutely no point in growing. There are also forms with variegated leaves, as well as a couple of varieties (notably the antique 'Brighton Rock') with irregularly striped flowers, which come true from seed. All these variations are just waiting to be bred into a range of plants of different growth habits and climatic adaptations.

Antirrhinum majus, the familiar garden snapdragon, is the source of essentially all varieties on the market today. It is grown as an annual, though it is technically a short-lived perennial in mild climates with excellent drainage, as in its native Mediterranean habitat; however, in most of the United States, anywhere with winters mild enough for it to survive will also have summers too hot for it to thrive. It is a happy self-sower and, if you don't deadhead, will frequently be a more or less permanent part of your garden. There are a handful of other species also native to Europe. *Antirrhinum braun-blanquetii*, which takes the prize for the strangest species name, has yellow flowers and significantly better winter hardiness (borderline for me in zone 5, but probably reliable in zone 6 and higher). *Antirrhinum hispanicum* and *A. molle* are low, mounded growers with pale pink flowers, fuzzy gray foliage, winter hardiness similar to *A. braun-blanquetii*, and excellent heat tolerance for southern summers. All the European species will hybridize, and the possibilities here are fascinating and varied. Commercially, some of these have been used to generate dwarf growth, but their real potential—for increased tolerance of heat and cold, novel flower forms and color combinations, and traits like the beautiful gray foliage—hasn't been explored.

Another group of *Antirrhinum* species native to the West Coast of the United States are so different from their European counterparts that some taxonomists place them in their own genus, *Sairocarpus*. As far as I know, all attempts to cross them with the European species have failed, but no one has explored the potential to breed *within* these West Coast species. There may be some interesting projects there for the adventurous breeder, if you can track these species down—none are commonly cultivated.

Snapdragon flowers are large and easily handled, and each flower produces many, many small seeds that are straightforward to germinate. I emasculate plump flower buds the day before they open. I find that if I firmly hold the base of the flower with one hand, pinch the base of the petals with the other, and gently pull, I can often remove all the petals and stamens in one quick, easy move. If that doesn't work (as it doesn't seem to with some individual plants) emasculation is very easy with a pair of tweezers. I've not seen bees visiting my snap flowers with the petals removed, so I don't bother to cover the flowers before or after pollination. All the European species except *Antirrhinum majus* are self-incompatible, so you can make crosses with them without even having to emasculate, though if you do so, you will need to exclude insect pollinators somehow if you have the right types of bees in your garden. The closed form of snapdragon flowers means that only the larger bee species, like bumble bees and carpenter bees, can successfully pollinate them. Honey bees are not able to pollinate snapdragon flowers, because they are not strong enough to pull the flower open.

If you try crosses between the species, the rule of thumb for snapdragons, and all species with self-incompatibility, is that crosses are almost always more successful if a self-compatible species is used as the female (seed) parent; in other words, you'll usually have more success using *Antirrhinum majus* as the seed parent and pollinating it with the other species, though I have occasionally been successful using *A. molle* as the seed parent in crosses with *A. majus*. In any case, the flowers are large and easy to cross, so I usually do crosses in both directions and see what happens.

Snapdragons are outcrossers with the typical inbreeding depression and F_1 hybrid varieties. As usual, embrace diversity in your populations. Snaps can also be propagated by cuttings if you have a cool greenhouse or a very sunny window where you can overwinter them, or if you are working with species and varieties that are perennial in your climate.

I started breeding snapdragons a few years ago, and it has been one of the projects that I've most enjoyed in my garden, filling it with tall, vividly colored spires of bloom that look gorgeous in a vase. It makes me laugh when I see the stunted little dwarf things they try to pass off as snapdragons at my local garden center. Start exploring this group, and you might just fall in love too.

ZINNIAS

Zinnias are one of the great neglected plants of modern horticulture. The current focus of the annuals market is on plants sold in full bloom in small pots in garden centers. Tall flowers, like classic cut flower zinnias, look lanky and awkward in a small pot, so they get passed over in favor of compact petunias and impatiens. The exception to this would be the newer hybrids with *Zinnia angustifolia* (the Profusion series is probably the best known), which boast mildew resistance and a small, compact growth habit that fits well with mainstream floriculture production systems; there is a lot of breeding work in this group, but the rest of the genus has been sadly overlooked. Luckily old-fashioned zinnias are easily grown from seed, and you can amass a huge collection of them for a few dollars. In the face of professional neglect, an enthusiastic group of amateur breeders is working with these plants, laying the groundwork for a new generation of amazing varieties.

Zinnias in gardens are mostly represented by three species and some hybrids between them. *Zinnia elegans* (formerly known as *Z. violacea*) is the classic, tall variety your grandma grew in her garden. The flowers are gorgeous and come in a dizzying range of colors and forms, though their Achilles' heel is mildew. Mildew is more ugly than lethal, but when it comes to flowers, ugly is not very acceptable. Despite that one flaw, they are still well worth growing.

Zinnia haageana has flowers in the color range familiar, primarily, from marigolds. Yellows, golds, oranges, and browns, usually in dramatic two-toned flowers. This species also tends to be more mildew-resistant, though it is still not totally immune.

Zinnia angustifolia (narrowleaf zinnia) is the new kid on the block, and commercial breeders have been having a field day with it. Imagine the scene at a kid's birthday party when the piñata breaks open and everyone goes wildly, joyfully scrambling for the candy. It's like that— only with *Z. angustifolia*, genes and breeders instead of candy and children. Lines derived from this species now dominate the market, and no doubt we'll see many more colors and forms of it in commerce in the coming years. But I rather think if you want something tall for the middle or back of the border, or something wonderful for cutting, you are going to have to breed it yourself.

All three of these species can hybridize. *Zinnia elegans* and *Z. haageana* cross fairly easily, and the hybrids are usually quite fertile. These hybrids are an exciting starting point for breeding, combining the bicolored petals and disease resistance of *Z. haageana* with the wide color range, large flowers, and tall plants of *Z. elegans*.

Zinnia elegans and *Z. angustifolia* are more difficult to cross. They have different numbers of chromosomes, so, like mules, the hybrids are sterile. The upshot is that this group is not very accessible for the home breeder, and unless you are really fascinated by the complexities of polyploidy, this cross is probably best left to the professionals. (The Profusion series and other commercial strains have had fertility restored by a chemical treatment that doubles their number of chromosomes.)

The possible goals for a breeder in this genus are many. Mildew resistance is, in most climates, an important priority, and any breeder should keep that in mind. Even if it isn't your primary focus, keeping a record of which individuals show the least disease and making it a factor in your selection is worthwhile.

Zinnias already come in nearly every color except true blue, but, especially with crosses between *Zinnia elegans* and *Z. haageana*, there are fascinating bicolors to be created. Many of the most interesting forms of zinnias are only offered in mixed colors, so there are a lot of quick and easy projects to select out the specific color or colors you like best from the mixes.

Flowers in *Zinnia* range from single to fully double and are of various shapes and forms. For example, the cactus types offer elegant rolled petals, and in the scabiosa-flowered form, the disk florets are slightly larger and more colorful but still contrast with the very showy ray florets. Backyard breeders have found that crosses between such diverse forms result in a range of new, exciting, and very beautiful flowers, rivaling even the intricacies of dahlias and roses. With more creative breeding work, I think we'll see this genus finally ready for its close-up.

Like all members of the daisy family, a zinnia flower is actually a cluster of many tiny flowers; that is, what we would normally call "petals" are actually each an individual flower, a ray floret, and the center of the flower is made up of the disk florets. In zinnias, the ray florets are exclusively female, while the disk florets have both male and female parts. To make a controlled cross, you can emasculate by simply removing the disk florets with tweezers, and then pollinating the ray florets with your male parent.

In double zinnias, most or all of the disk florets have been converted into ray florets, giving a full look to the flowers. Fully double zinnias have only ray florets and so don't require emasculation. Unfortunately, since they don't produce pollen, doubles must be crossed with single or semidouble forms, making it all but impossible to create a uniformly double seed strain.

The stigmas have a forked, Y shape. When immature, they are held together, and when they spread apart, they are ready for pollen. Collect some of your male parent's pollen with a tweezer, or simply pick the entire flower, dust it over the stigma, and you are done. Zinnias are insect-pollinated, so to be sure other pollen doesn't get in, it is best to cover immature flowers before pollination and for a day or two afterward.

Zinnias are outcrossers, so commercial varieties are either diverse populations showing a lot of variability, or F_1 hybrids.

Classic zinnias are one of the easiest flowers to grow, widely adaptable, tough, and simple to cross. If you are looking for a great plant to start with, or a project to do with your kids, this may be just the group of plants for you.

BEANS

I never used to grow beans. Beans, after all, are beans, right? Why invest the garden space in them when they are so cheap at the store? As for green beans, well, I always thought I didn't really like them. But on a whim, I tried a few varieties, and that started the beginning of a collection. Get into heirloom dry beans, and you'll find a mesmerizing diversity of beautiful colors and patterns and an equally wide range of flavors and textures, quite unlike the selections available from your average grocer. Even better, all freshly harvested beans cook up quickly, with a wonderful smooth, creamy texture, compared to the stale bags that have been sitting around on a store shelf for goodness knows how long. Fresh-from-the-garden green beans were also a revelation, and some of the intensely flavorful Italian heirloom varieties changed me from not really liking green beans to considering them one of my favorite vegetables.

Dry beans suffer seriously from a lack of breeding for flavor and eating quality. If you buy a bag of black beans or pinto beans at the grocery store, you are not buying a variety like you do when you pick up a 'Honeycrisp' apple. Rather, they are market classes, meaning any variety that produces small, dark beans gets sold as a black bean, while large, speckled varieties get sold as pinto. The result is farmers are forced to compete solely on how cheaply they can get varieties to market, so commercial breeders focus exclusively on traits like yield and disease resistance. Flavor is such a nonissue that one very successful professional bean breeder I talked to told me that at no point in the selection process does he ever taste his varieties. If you want tough, high-yielding varieties in a small range of colors, commercial breeding has got it. Get your hands on a good heirloom seed catalog, however, and you'll find a deluge of incredibly beautiful, delicious beans. New colors and patterns are waiting to be discovered, along with even more exceptional flavors and textures for specific uses. Once you've explored the diversity out there, you'll find an endless array of projects to create even better flavors, or adapt the ones you like best to grow better in your particular climate.

Green beans suffer from a slightly different problem. The very best green beans for eating are tender and fiberless; unfortunately, the very best green beans for grocery stores are firmer, so they ship well and don't wilt in the produce bin, meaning commercial breeding has actually been focused on creating tougher varieties. Revel in the luxury of eating what can only be harvested in your own backyard and indulge in delicate, gourmet-quality green beans.

Breeding for flavor in green beans has been focused almost exclusively on varieties that are "mild." Some people like it, but I find it dull, especially compared with the rich, wonderful flavor of heirloom Italian varieties like 'Romanesco'. There is a surprising diversity of flavors to be found in green beans, some of which you may like and others you may not. The safe, most widely marketable strategy is simply to breed against all strong flavors, so commercial varieties tend to be bland. I hope you'll ignore that, taste through a range of delicious heirlooms, and start breeding for the flavors you enjoy most.

Expect a little bit of a learning curve when making bean crosses. Once you get the hang of it, they're actually extremely quick and easy to cross, and if you do it right, don't even require emasculation. But their flowers are complex and unusual. To get familiar with the parts, pick an open flower from your plants and look at it. The back of the flower, at the top, is a broad petal, called the standard. In the very center is the keel, an odd, folded-over petal that actually twists in a corkscrew. Hanging down on either side of the keel are two little petals called the wings. The anthers and stigma are inside the corkscrewed keel, twisted around each other, with the anthers sitting right on the stigma, ready to self-pollinate it.

Here is the cool part. Grab those two little wing petals and pull them gently down while holding the rest of the flower still. As you pull down, poking out of the tip of the keel will come the stigma, slowly spiraling out as you pull the petals down. Push those two petals back up, and it will retract. Cool, right? On a fully open flower, when the stigma comes out, it will already be covered with pollen (which you can see as a fine powder if you look closely), meaning it is too late to pollinate it. On a slightly immature flower, it will come out pollen-free and ready to be pollinated.

To make a controlled cross, you have to get at the buds first thing in the morning on the day before they are going to open. In bud, the standard petal will be wrapped around everything else, forming a neat, smooth package. Buds ready to open will be plump and full, and if purple when fully open, just beginning to show streaks of color. With a pair of tweezers, carefully peel back the standard petal, then gently pull down on the little wing petals to pop out the stigma. To pollinate it, grab a fully open flower from the male parent and again pull down its wings to get at the stigma. That stigma will be covered with pollen, so just grab it with your tweezers, and use it to dab a bit of pollen onto the tip of the other flower. If you want, you can hook the curves of the two stigmas together to keep them in contact. Some people like to then tuck the wing petals back in and wrap the standard petal back over everything. The benefit is to keep the humidity high around the flower, which is said to help the pollination be successful. In my moist climate, this doesn't seem to help, so I never bother, but if you live somewhere hot and dry, it might be worth

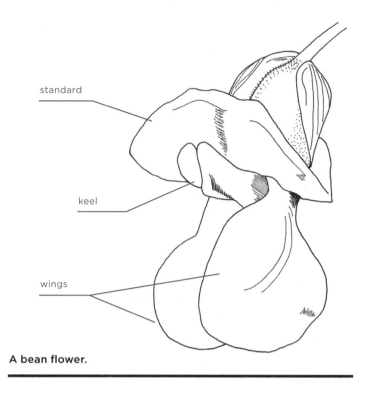

standard

keel

wings

A bean flower.

---

trying. Because beans are such confirmed selfers, there is no need to cover the flowers to keep pollinators out, and because you are getting in with the pollen you want before the self-pollen is released, there is no need to emasculate either. Just pollinate, label, and go on your merry way!

Beans are self-pollinating, showing essentially no inbreeding depression. The flowers are built in such a way that, unless you (or *very* occasionally a bee or other insect) intervene, the pollen from the maturing anthers is released directly onto the stigma to instantly self-pollinate. This means that usually, even without any efforts to isolate your plants, you can be confident that the seeds you harvest are self-pollinations—though, as with every-thing, the frequency of crossing depends on your region.

Carpenter bees are particularly effective at cross-pollinating bean flowers, so if you have a lot of them in your area, it might be worth covering flowers for varieties you particularly care about to ensure they self-pollinate.

The wild ancestor of the common bean was domesticated twice, simultaneously, in pre-Columbian America. These two parallel domestication processes have resulted in two groups of beans: the Andean bean from the mountains of South America and the Mesoamerican bean from Mexico. In general, Andean beans are larger (think kidney and pinto beans), while Mesoamerican beans are smaller (think black beans), though exceptions to that broad-stroke rule are numerous. What's important to know is that the groups differ genetically, so much so that though they can be crossed, hybrids between the two are frequently lacking in vigor and have reduced fertility. So if you make a cross, and the seedlings just don't want to grow or set seed, that may be the problem, and you might be better off attempting a different cross to reach your goals.

The common bean is the species *Phaseolus vulgaris*, and though lima, adzuki, garbanzo, and soy beans are also called beans, they are all in completely different genera and (though fascinating to breed in their own right) can't be hybridized with common beans. The scarlet runner bean, however, is a different species in the same genus, *P. coccineus*, and it can cross with common beans, with difficulty. As with most wide crosses, numbers are your friend here; expect to make a lot of crosses to get a successful hybrid, and some varieties cross more easily than others. Though not easy to achieve, the results of this sort of wide cross might be interesting. *Phaseolus coccineus* isn't as tasty (in my opinion) as common beans, but it is a lovely plant, perfect for the ornamental edible garden, and grows better in cooler temperatures. Hybrids between the two species could, potentially, produce more delicious, even more beautiful beans for the home garden, and could be a good place to go for gardeners in cool summer climates.

Common beans come in two different growth habits, a trait controlled by a single gene: indeterminate or pole beans, which produce long vines that continue to grow and flower all season, right up to frost, and determinate or bush types, which produce vines that grow to a certain length and then stop. Most determinate types are selected to be very compact bushes, but some will produce a short vine.

Indeterminate types start flowering later in the season, but keep on producing over a much longer period and do require something to climb up. Determinate types begin producing earlier and don't need to be staked but also stop yielding once they stop growing. In northern climates with short summers, determinate beans will usually have higher overall yield, while in places with a longer growing season, indeterminate types will win out. Personally, I grow all bush beans for my dry beans, and a combination of bush and pole beans for green beans, the bush beans providing my early yield and the pole beans producing at the end of summer, when the bushes have ended their run.

For green beans, remember that it takes some time for the pods to ripen and mature, so harvest, taste, and select early in the season, then leave the pods of your preferred individuals on the plant to mature. Unpicked green beans, like unpicked zucchini, effectively shut down the production of new flowers and pods, so you won't be able to continue to harvest from the plants you want to save seed from.

For dry beans, be sure to put your harvest of beans from any one plant into a separate bag so you can keep them organized, and cook up and taste a sample from each variety. Cooking and tasting a large number of different bean individuals can be tedious. The most effective method I've found is to presoak a sample of beans from each individual, then place them in the cups of a muffin pan. Fill each cup with boiling water, add salt, cover the whole thing tightly with aluminum foil, and place in a 375° Fahrenheit oven. They'll be done in a little over an hour, less if they are very freshly harvested. If you are interested in selecting for quick-cooking varieties, you can check them periodically to see which varieties cook the fastest.

For ease of culture, deliciousness, nutrition, and sheer collectability, it is hard to beat beans. Go and start having fun making new ones!

CABBAGE AND OTHER BRASSICAS

Cabbage, broccoli, kale, cauliflower, kohlrabi, collards, ornamental cabbage, Brussels sprouts—all these apparently distinct vegetables get lumped together here because they are all simply different versions of the same species, *Brassica oleracea*, so what is true of breeding one is largely true of them all. It is one of those facts that blew me away when I first learned it and amazes me still. Corn wins, in my mind, as the crop that least resembles the wild plant it came from, but I think *B. oleracea* stands alone in terms of the sheer diversity of the cultivated forms it has morphed into. Humans have bred this one species specifically for eating its leaves (cabbage, kale, collards), stems (kohlrabi), buds (Brussels sprouts), and flowers (broccoli and cauliflower). Basically the only thing we haven't bred this one for is eating the roots—though you can eat the roots of the very closely related turnips and rutabaga, and since rutabaga is the result of a hybrid between turnips and kohlrabi, it almost counts. The seedlings can be eaten as sprouts, various ornamental cabbages and kales are grown to add interest to fall and winter gardens, and if you want to get *really* crazy, there is a fun heirloom variety actually bred to grow a very long, woody stem that can be used as a walking stick.

This incredible diversity makes me confident that there are a lot of wonderful things yet to be teased out from this group of plants. As usual, professional breeders tend to stick to the familiar, established forms, which translates to lots and lots of room for us to get creative and blaze new trails.

The brassicas are outcrossers. The small, yellow flowers are pollinated by insects, so it is impractical to isolate them by distance in the average garden. Luckily, even though this species is common in gardens, it (in all its varieties) is not typically allowed to flower, so as long as you have only one population at a time flowering in your own garden, you can generally rest easy, knowing that you won't be getting pollen carried in from your neighbors' gardens or a farmer's field. Brassica flowers are also self-incompatible, which means making crosses is quite easy. As long as you cover flowers before they open with some kind of netting to keep the insects out, there is no need to emasculate. Just dust them with the pollen of the other parent of your cross, and the seeds will all (or nearly so) be hybrids. If for some reason you want to self-pollinate, you can get around the incompatibility if you make bud pollinations.

The most difficult thing about making brassica crosses (at least in my zone 5 climate) is not the pollination itself but getting the plants to flower. With the exception of broccoli, these vegetables are all biennials, meaning the first year in the garden, they just grow leaves and don't switch to flowers until they go through the cool temperatures of winter. In the garden this is a good thing—I don't want my kale or cabbage flowering, because if they flower they stop growing leaves, and the leaves are what I want to eat. In warmer zones (about zone 7, in a sheltered location, and higher) this isn't too much of a problem: simply leave your plants in the ground over winter, and the next spring they will flower. For cabbages, you might need to slice the head open so the flower stalk can emerge from the closely packed leaves, but that's it.

But in my chilly world of Michigan, it isn't so easy. Leave the plants in the ground, and they will die, unless I have a particularly mild winter. I've tried a couple of different techniques to get around it. One is to grow plants in the shelter of a cold frame, which usually is enough to get them through the chill. I've also had fairly good success digging up the plants, washing the soil from their roots, storing them in the refrigerator over winter, and planting them out again in the spring. The fridge keeps them cool enough to give them the chilling they need to flower. They look pretty rough come spring, but I get a few flowers out of them, which is all I really need. To make my own life easier, I've started leaving as many plants as possible outdoors over winter and saving seeds from anything that survives. Winter hardiness doesn't really impact their value as a vegetable, but it sure makes breeding easier. Now and again, an individual will flower its first year, but I pull out those plants right away and never save seeds from them: first-year flowering means less leaves to harvest, which, for me, means it loses points as a vegetable.

You could go a million directions with brassica breeding, but I think several options are particularly worth pursuing for the home gardener. Broccoli, for example, typically produces one giant head and that's it. This is perfect for commercial growers, who want to harvest a field once with a machine, but in the home garden, I like my harvest spread out a little. Some varieties produce smaller heads and, after harvest, will send up a second round of flowers to harvest again. I'd love to push this to the point where I can plant once and keep harvesting the entire season. One interesting twist on this (for gardeners in warmer climates than mine) is a perennial version that can keep producing year after year. More flavors and colors and such would be fun to add to this perennial-type growth habit.

The other exciting area for exploration is the fact that since all these vegetables are the same species, they are all easily crossed. I've been seeing a lovely broccoli produced by crossing regular broccoli with an ornamental cabbage in catalogs recently; it is gorgeous, with amazing purple stems. A similar hybrid of kale and Brussels sprouts, trademarked as Flower Sprouts, is just becoming available. There are many other fascinating options here. A kale-broccoli hybrid you harvest leaves from the first year, and then broccoli the second year, after it overwinters? Or, more practically, fussy cauliflowers that never want to grow for me getting a bit beefed up and more vigorous, thanks to crossing with tough-as-nails kale? The possibilities are endless.

If you really want to get adventurous, *Brassica oleracea* can hybridize—with difficulty—with two closely related species, *B. rapa* (turnips and a whole range of delicious and beautiful Asian greens, from mizuna to bok choy) and *B. nigra* (black mustard). In fact, *B. napus* (rutabaga, canola) and *B. carinata* (Ethiopian mustard) derive from long-ago crosses of *B. oleracea* with *B. rapa* and *B. nigra*, respectively. Recreating these crosses yourself will not be easy and requires bud pollination to be successful at all, but there is interesting potential. The leafy green varieties of *B. rapa*, especially, could be absolutely fascinating, and lovely, combined with kales.

When I look at the range of beautiful, incredibly nutritious, and delicious varieties that have already been created from simple *Brassica oleracea*, I get excited, wondering what else lies within this one remarkably diverse species.

CORN

I find myself endlessly fascinated by corn, the basis of almost all the food we eat. Corn is among the plants most radically transformed through history by human breeding, uniquely mutable and able to change under selection. Traditional native American varieties include a huge amount of diversity, virtually all of which has been lost or relegated to use as decoration, while commercial breeding focuses almost exclusively on yellow dent corn for animal feed, plus a tiny bit of sweet corn. The disconnect between corn's history, potential, and current use, tells me that if there is anywhere for home breeders to dream big, get creative, and make something wonderful, it is with corn.

Corn is an outcrosser. Practically the poster child of outcrossing plants, in fact. Of all the plants in the garden, this is one where it is very important to avoid inbreeding if you want to maintain vigorous plants and good yields.

Corn has a very unusual structure to its flowers. The silks on the ears are actually the female parts of the flowers, and the tassels at the tops of the plants are the male portions. Pollen is released by the tassels, and blows around on the wind to pollinate the silks. The flower structure makes controlling pollinations pretty easy. You can emasculate a plant simply by grabbing the tassels when they start to emerge and pulling them off (they come out with a very satisfying POP! sound). If you have just two varieties you want to cross, simply plant them in rows next to each other, detassel one of them, and all the seeds of your detasseled plants will be hybrids. The only tricky part is making sure that the two plants come into flower at the same time. Commercial breeders will make a series of plantings over a period of a couple weeks in the spring, so they have plants at all stages of maturity for making crosses. In my experience, this usually isn't necessary, at least in my garden. I plant all my corn at once, and find that the silks usually stay receptive long enough for me to make the crosses I want, even though things mature at slightly different times.

If you want to make multiple crosses, you either need to isolate your different plantings in different parts of the garden to keep them from cross-pollinating, or make the crosses by hand. Since corn is pollinated by the wind instead of bees or hummingbirds, it is actually quite easy to give them enough space to ensure they won't cross-pollinate. This may seem counterintuitive, but unless you have a very windy garden, you should not expect corn pollen to blow more than a dozen feet from the plant. Bees, on the other hand, routinely fly for a mile or more with pollen. The commercial standard for isolating plants is 120 feet. At that distance, there is essentially no chance of pollen from one plant getting to the other. In my experience, a dozen feet is usually enough if you don't mind a few cross-pollinated seeds here and there, particularly if there is something tall in between to block the wind. A couple rows of sunflowers works well and is pretty to boot.

Hand pollination is pretty straightforward as well. The standard procedure is to take a paper bag (there are special bags sold specifically for this purpose, though a simple paper bag works fine as long as you don't have a rainy summer) and slide it over the developing ear before the silks emerge. A couple days later, once silks are showing, tassels from the male parent are pulled off, placed in another bag, and set in a dry spot. The pollen will fall off and collect in the bag, then you go out and dust the pollen over the silks, putting the bag back over it to protect it from other pollen. A few days later, you pull the bag off and label the developing ear (permanent marker on the husk of the ear is an easy way). Personally, I never have paper bags around, and my summers are very rainy, so I use a different method. I take a strip of aluminum foil, and form it into a loose cap over the ear to protect the silks, and then I just take an entire tassel, place it directly on top of the silks, and kind of crimp it in place with the foil. Unorthodox, but it has worked fine for me.

One of the most interesting things about working with corn is that the seed itself can give you a lot of information. When you pollinate a tomato, what you are harvesting and eating is fruit produced by the mother plant. No matter what you pollinate it with, the fruit will taste the same. To see the results of the cross, you have to wait until you grow the next generation. With corn, however, you are looking at the seed itself, and its color and shape is the result of the genetics of that individual seed. So if you pollinated a white corn with a mixture of its own pollen and pollen from a blue corn, the ear will be a mixture of blue and white kernels. The white ones will be self-pollinations, while the blue ones are crosses with the blue corn. This is very useful to detect cross-pollination between populations of corn, and also makes it possible to make selections for many traits before you even plant the seeds.

With so much beauty, history, and diversity in corn that has been all but eliminated by the tide of commercial corn farming, I think it is high time we gardeners began restoring this lovely, infinitely variable plant to its former glory. Whether you pop it, eat it while sweet, or use it to decorate your fall table, this unique plant will amply repay any time you spend with it. Each of the many uses of corn will require different breeding practices. For any of them, exploring the wide diversity of corn to create increasingly delicious, beautiful varieties will be very fruitful.

Ornamental corns

The multicolored Indian corn we see each fall is a fascinating area just waiting to be explored. From short stubby ears that fit in your palm to ones over a foot long, with kernels of red, white, yellow, blue, purple, green, and black, the sky's the limit. Since the color of each seed reflects its individual genetics, sorting the mix of these standard decorative flint corns into specific color strains or combinations of colors is almost effortless. Want black and orange ears by the lamppost for Halloween? Or corn in the colors of your favorite football team for the homecoming tailgate party? It could hardly be easier. Buy an ear (or packet), pick out the colors you want, and plant. It isn't always an instant success. Remember, a recessive color might be hiding behind a dominant color; and my efforts to breed a green-and-white corn to celebrate the local team at Michigan State have been frustrated by the fact that green is a combination of blue and yellow pigments, and when crossed with white, they segregate independently, so my corn is white and green . . . and blue and yellow . . . Not so good, given blue and yellow are the colors of our sworn enemy, University of Michigan. But I'm having fun, and I can still grow separate green and white ears and display them together, even if I haven't yet figured out how to get the two colors together on the same ear.

Popcorn

I am a huge of fan of popcorn, and there are lots of fun things to do with it. Flavor is, of course, a big focus for me, and there are several strains of gourmet popcorns I'm working on combining for truly exceptional flavor. I dreamed, for a moment, of multicolored popcorns, but the problem is when you pop corn, what you are seeing is the puffed inner starch, and the color of corn is only in the outer layer, so whatever color the corn, it all pops white. But beautiful multicolored popcorn ears do great double duty as seasonal decorations, whose kernels can then be stripped off their cob, popped, and eaten.

Sweet corn

Breeding sweet corn is great because, just like with color, you can see it in the seeds before you plant them. A sweet corn kernel will be shriveled and wrinkled-looking when mature, and the sweet genes are all recessive, so when you see a shriveled kernel, you *know* it is homozygous for the sweet genes and will produce uniformly sweet off-spring. There are three genes involved in making corn sweet. The *su* gene is the original, old-fashioned sweet gene. The *se* is the sugary enhanced gene, and when it is present in combination with the *su* gene, the kernels are sweeter and stay sweet longer. Finally, there is *sh2*, or the supersweet gene.

The first thing anyone thinks to do when breeding sweet corn is to cross it with colored ornamental corn. Conceptually, this is super easy to do, because you can select right on the ear for sweet, colored seeds. The problem is, most colors in corn don't develop until the seed is mature, and sweet corn is eaten, of course, when immature. There are varieties, like 'Inca Rainbow', which have been bred for multiple colors, but at the eating stage, they'll be so pale you'll hardly notice them.

Flavor is, of course, another matter—there's a ton of potential here. Commercial breeding is focused pretty much exclusively on yield and sugar content. The more complex nuances of flavor remain for the home gardeners to explore and create.

One challenge of doing taste tests of sweet corn is that, of course, you have to eat it before it is mature, so if you have only one ear on a plant, you can either see what it tastes like, *or* let it mature for seed, but not both. My solution, which works most of the time, is to plant with a wider than usual spacing. Give a couple feet between each plant, and each one will almost always produce two or more ears so I can taste one, and then decide which plants to save seed from.

Corn for flour and meal

Ground corn, cooked into grits, polenta, or cornbread, is delicious and well worth growing at home, if you buy a mill to grind it with. Besides the flavor of different varieties, fresh-ground, whole-grain cornmeal of any variety is orders of magnitude better-tasting than the stale, degermed stuff you get at the grocery store. When it comes to grinding corn, there are three basic types. Flour corns have large kernels with a soft starch that is easily ground into a fine flour. Flint corns have a different type of much harder starch, which is usually ground into coarse cornmeal. Dent corn, a cross between the two types, contains both types of starches and falls somewhere in between.

Nixtamalization

Before Europeans arrived, essentially all corn was prepared by nixtamalization, a process in which the corn is cooked with lye (calcium hydroxide, often sold as pickling lime or red lime paste at Asian grocery stores). This softens and removes the outer coat and makes the starch soft, easier to digest, and more nutritious. It also produces the characteristic "corny" flavor of traditional corn tortillas and tamales. One of the best things about this process is it doesn't require a special mill like you need to produce cornmeal. Simply simmer the corn with the lye, rinse it, and then it can be puréed in a food processor and formed into the most amazing tortillas and tamales you've ever had. It isn't that complex a process, and if you love authentic Mexican food, or are interested in traditional food ways, it is a practice well worth exploring.

LETTUCE

Lettuce is one of those vegetables everyone likes. So basic and yet so good, and quite fun to grow because you can get such a range of delicious and beautiful colors and shapes and frills and textures, from looseleaf to icebergs and other tight heading lettuces. And no matter how good your local grocery store is, nothing can quite beat the crispness of lettuce just picked with the morning dew still on it. Lettuce can also be a bit frustrating. When grown in the cool weather of spring and fall, it is tender and delicious, but with the hot, longer days of summer it quickly turns bitter and strong-flavored—and then bolts. Instead of growing as a low, compact rosette of leaves, the stem rapidly elongates and produces tiny, yellowish flowers and a crop of small, fluff-bedecked seeds, rather similar to dandelions, to which lettuce is actually not too distantly related.

Lettuce also has an underappreciated role as an ornamental. Leaf color ranges from a bright chartreuse to rich, deep red, and the leaf shape can be long and slender or exuberantly frilled. Grown as an ornamental, bolting is less of the problem, with the bitter leaves actually looking quite dapper arranged in a neat spiral around the long stem. The flowers, however, add nothing ornamental to the display, and shortly after flowering, the whole plant dies an ugly death.

Lettuce and other gourmet greens have been having their day in the sun lately, and any good vegetable seed catalog will offer a mind-boggling array of different varieties. The directions you'll want to go in breeding new ones will depend on how you like to grow and eat your lettuces. I like looseleaf lettuces that I can plant once and harvest many times over a fairly long period of time, so slow-bolting varieties are a high priority for me. If you prefer heading lettuces, where you harvest the entire plant and then sow a new crop, bolting will be less of an issue as long as it isn't so early as to interfere with head development. Similar individual tastes will rule when it comes to flavor and texture. Do you like thin, delicate leaves? Thick, meaty, spinach-like ones? Crisp, crunchy heads? Lettuce can be all these things. There are even varieties, called celtuce, which are bred for their thick, edible stems. And then, of course, there is the visual effect of all those colors and frills and shapes. With all the variables going into a perfect lettuce variety, you can easily spend a wonderful time simply combining all the things you like best into a few splendid varieties, or pushing the boundaries with later bolting or exceptional flavor and texture.

Cultivated lettuce is all derived from one species, *Lactuca sativa*, and it crosses easily with three other wild relatives, *L. serriola*, *L. saligna*, and *L. virosa*. You may have actually seen a couple of these plants without realizing it, as *L. serriola* grows as a weed across the United States. Once I learned what it was, I realized I'd seen it growing around in vacant lots and litter-filled alleyways. These related species are not really palatable, being extremely bitter and spiny, but they might have interesting diversity to bring to lettuce. These wild species have been used in commercial breeding, a little, to provide increased disease resistance, but the potential for new leaf forms and flavors from these wide species crosses hasn't been explored. Also potentially useful: some forms of *L. serriola* grow as biennials rather than annuals, meaning they don't bolt to seed their first year, and might be useful in breeding super-slow bolting forms of lettuce. Working with these species might be interesting but will also be a long row to hoe in terms of getting the flavor back to something better than merely edible.

Lettuce is a selfer, which makes selecting new varieties easy, but creating new hybrids is a bit difficult. The incredibly tiny flowers can be emasculated only by wielding the most finely tipped tweezers with the steadiest possible hand. I simply don't have the manual dexterity to do it, though some people can. There are a couple of ways to skirt the necessity of emasculation. The main method used by professional breeders takes advantage of how the flowers develop. The pollen is released first, covering the inside of the flower, before the stigma is receptive, so there is a window in which a well-timed spray of water can wash away all the pollen, leaving the flower clean and ready to receive the pollen you apply to it.

To get the hang of it, you'll have to experiment. Examine developing flowers a couple of times a day, and try rinsing them down with a firm spray of water with a small squirt bottle at different stages of development, ideally as soon as possible after you can see the powdery pollen on the flower. Don't pollinate them, and if you've hit upon the right stage for rinsing, you won't get any seed set from those flowers. Once you've found the right stage for rinsing, simply spray down the flower, let it dry, then apply your pollen. Even with experience, you're likely to get a fair amount of self-pollination. When possible, try making crosses between varieties with a lot of visible differences, so you'll be able to pick out the self-pollinations. For example, a cross between a dark red and a green variety will produce intermediate-colored seedlings that will be easy to distinguish from the leaf color of the parents.

For a lazier, lower-key approach, you can just plant the varieties you want to cross next to each other and tie their developing flower heads together to increase the chance that some pollen will move from one to the other. Most of your seedlings will still be selfs, but provided you can visually pick out the crosses by leaf shape or color, you'll be able to find your hybrids and start your breeding program. After all, all the F_1s will be the same, so you really need only one that lives to maturity and produces F_2 seed to select from.

Despite the difficulties in making lettuce crosses, they're a great crop to work with if you don't have much space in the garden, or are impatient. Individual plants are small, and you can evaluate traits like leaf shape and color very soon after germination. Even for traits like bolting and heading, it takes as little as a couple months to evaluate, so you can easily grow two generations in a year, and very rapidly create and finish something new and beautiful. Jump that one little hurdle of perfecting your crossing technique, and you'll have a whole amazing world of diversity to play with.

SQUASH

This has got to be the perfect genus of vegetables for the home breeder: delicious flavors, both immature (as summer squash) and mature (as winter squash and pumpkins); incredibly gorgeous fruits in a bewildering range of colors and sizes; flowers that make crossing super easy without emasculation; minimal inbreeding depression; and big seeds, which can be sown directly in the garden without any fuss about starting seedlings indoors. The only downside to squash is that they can be big and take up a lot of space in the garden, but you can get around this with bush varieties (bush growth habit is conferred by a single recessive gene, *bu*, making it easy to add to new varieties) or growing the long vines up trellises.

The few familiar forms (green zucchini, pumpkins, butternuts, acorns) only scratch the surface of the diversity to be found in this group. A great multi-use plant, you can harvest summer squash all season as a green vegetable, while winter squash are easy to store, keeping you in homegrown produce until spring, when the garden starts

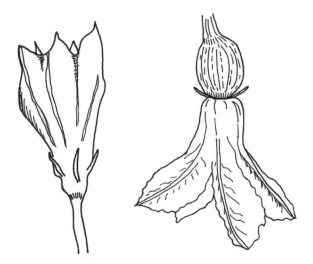

Male (left) and female (right) squash blossoms.

up again outside. Along the way, winter squash are lovely and can be decoratively displayed, indoors or out, right up to the moment you decide to eat them.

Despite the diversity within squashes, commercial breeding is focused almost exclusively on the familiar types just named and practical values like yield and resistance to disease. The disease-resistance work is very welcome, given the propensity of many squashes to mildew, and I welcome the increases in yield in winter squashes (but I rather think they've gone a bit overboard on the zucchini). But, as usual, flavor and beauty have been—with few exceptions—neglected.

Squashes produce separate male and female flowers. The two are easy to tell apart. The male flowers are produced first and in much greater numbers. You can recognize them in the bud because they sit on a smooth, straight stem. Female flowers, on the other hand, will have a tiny, baby version of a squash sitting just at the base of the flower. Every morning, a new group of flowers open and are instantly mobbed by bees, wasps, and other pollinators. Because insects can and do fly a mile or more to flowers, you have to assume that any open-pollinated flower has been visited by pollen from any squash in the neighborhood.

Making controlled crosses is incredibly easy, however. In the evening, visit the squash patch and search for buds that look like they will open the next morning. They are easily recognized—the bud will be large and flushing yellow, in contrast to the smaller, green buds that still have time to mature. Take a bit of tape (or string), and tape (or tie) the tip of the male and female buds you want to cross, to keep them closed. Go out first thing the next morning; you'll see other buds have opened into flowers, but the ones you taped will still be held closed, keeping the bees out. First pick the male flower, then go to the female and pull the tape off it. Pull the petals off the male, and smear the pollen all over the female stigma. Do this fast, as bees can slip in surprisingly quickly with other pollen when you aren't looking. Once you've pollinated, tape the female flower closed again with a fresh piece of tape. The flower will shrivel and fade by that afternoon, and if the cross was successful, you'll see the fruit begin to swell and develop quickly.

If you are breeding winter squashes, you can let the fruit mature, and store and eat them as you usually would, simply pulling the seeds out before you cook it. Summer squashes like zucchini are eaten immature, so you have to let the pollinated fruits develop fully, which, unfortunately, tends to shut down the production of new zucchini on that plant. You might be tempted to harvest for a while before making crosses, but don't wait too long. Squash takes a long time to ripen and mature, so in northern climates you pretty much need the whole summer for them to develop, and in most areas, mildew and vine borers tend to knock plants down by the end of the summer, so the best rule is to pollinate as soon as you can. When I select zucchini, I harvest one from each plant, so I can evaluate flavor, and then pollinate my favorites. When I'm breeding for things like resistance to mildew or borers, which don't show themselves until later in the season, I pollinate everything and then save seeds only from the plants that performed the best.

Cucurbita pepo is the most common of the four species typically encountered in gardens. Almost all the summer squashes (zucchini, crook necks, patty pans) are pepos, along with popular winter squashes like the spaghetti, delicata, acorns, and most of what we call pumpkins. In addition, most of the small decorative gourds are pepo varieties. In general, pepo varieties don't store very well and tend to be susceptible to mildew and vine borers.

Cucurbita moschata is best known as represented by the butternut winter squashes, though there are other varieties, and one ('Tromboncino') is grown as a summer squash. Moschatas aren't as diverse as the pepos, but their flavor is wonderful, and they are notably disease- and borer-resistant.

Varieties of *Cucurbita maxima* are the queens of the winter squashes—incredibly beautiful, with unsurpassed flavor and a very long storage life. As the name "maxima" implies, these can be incredibly massive. The giant pumpkins seen at fairs and holders of world records are varieties of *C. maxima*.

Cucurbita mixta (to some, now *C. argyrosperma*) is probably the least grown of the common species. Mixtas include some very nice winter squashes, including some attractive pumpkins. They are particularly well adapted to hot climates, and like moschatas, they are generally disease- and borer-resistant.

These four species can be hybridized for fascinating results. As a general rule, *Cucurbita moschata* can cross with *C. pepo*, *C. maxima*, and *C. mixta*, but the other three are less able to cross with each other, though *C. pepo* has been successfully crossed with *C. maxima*. These types of crosses are not a guaranteed thing. How successful the cross will be depends greatly on the individual varieties being used. Some cross fairly easily, others not at all, and still others will cross but be completely sterile.

Commercial breeders have been making these crosses for a long time, but as usual they have their eyes trained on pleasing the mainstream market. Their no-nonsense focus has been simply to move over disease-resistance genes but keep everything else the same. If they cross a zucchini with a moschata, they'll then carefully cross back to a zucchini, so that the final product looks and tastes just like the zucchini you are used to, but has a couple of genes that help it resist mildew from the moschata parent. Practical, of course, but where is the fun in that? If I'm going to go to all the effort of crossing two completely different species, I want to select out something entirely new . . . new tastes, new looks, all the fun stuff commercial breeders don't get a chance to play with.

TOMATOES

Tomatoes are the quintessential home garden vegetable. Even people who don't consider themselves gardeners often grow a plant or two in a container. The sheer number of varieties—colors, flavors, growth habits, shapes, forms—available in catalogs is mind-boggling. Somewhat surprisingly, though, all that visual diversity may be traced to differences in only a few genes, and heirloom tomatoes are, on the whole, very genetically uniform. Modern commercial breeding has worked hard to change that, using wild relatives from South America to improve disease resistance and yield and create a more robust, diverse set of varieties that, counterintuitively, all look very much the same. The potential for a new generation of heirlooms combining the beauty and flavor of old varieties with the vigor and diversity of the new is huge.

With tomatoes, flavor rules—and discussion and opinions abound (they are right up there with favorite wine grapes and apples for topics to avoid in polite conversation). The flavor of tomato is complex, with many, many factors going into it, but there are several dominant elements to the taste of a tomato, and it is worth thinking about each of them individually as you taste and select your tomatoes.

I love an intensely sweet tomato, but not everyone does. Sweetness is quite easy to evaluate with your tongue, though acidity can interfere with your perception of it (a tart tomato will taste less sweet than a less acidic one with the same amount of sugar). If you want to get serious about it, you can actually buy a cheap "brix" meter, which will give you an exact measurement of sugar levels. Incidentally, because sugar is so easily quantified, commercial breeders have done more work with this, producing some intensely sweet (but to my tongue, at least, otherwise quite bland) varieties.

Acidity, like sweetness, is easily evaluated with your tongue, though you can also measure it precisely with a pH meter if you want to get obsessive about it. Personal tastes vary widely here. Do some tastings, and find out what you prefer. I like very tart, very sweet tomatoes for eating fresh, but for a pasta sauce, I find I prefer a little less acidity. Play around, and see what you like best.

Umami is a slightly mysterious taste which, along with sweet, sour (acidic, tart), bitter, and salt, is actually detected by dedicated taste buds on your tongue. The flavor is caused by glutamates, which generally form when amino acids in proteins break down in cooking. Thus, umami is a major aspect of the rich, full, satisfying flavor of meats and broths. It is also highly concentrated in classic flavor enhancers such as parmesan cheese, soy sauce, and the modern version, monosodium glutamate. Tomatoes, almost uniquely in fresh vegetables, also have a big hit of umami, and that is a big part of what makes them so delicious.

Umami is hard to describe. It is a flavor we are seldom directly aware of—often, it seems to simply make all foods taste better. But since it is a huge part of why tomatoes are such a wildly popular vegetable, it is well worth taking the time to train yourself to detect it. To my tongue, the best example of umami is offered by the variety 'Black Krim'. Take a bite, and you'll notice it is a quite mild tomato—not particularly sweet or tart, and yet it has a rich, full, lingering character that is hard to put your finger on. That is the umami. If you have an Asian grocery store in your area, you can also train yourself to detect the flavor by getting a bag of monosodium glutamate (often labeled as "aji no moto"). Take two slices of a mild tomato (such as a flavorless grocery store variety) and sprinkle a little (just a little, it is very concentrated) on one slice. Then take a bite of each. That rich, fuller, hard-to-define "moreish" betterness of the treated slices is umami. Pay attention to it in your breeding, and you'll get astonishingly delicious tomatoes.

Other aspects of flavor are detected by your nose. The compounds there are too numerous to mention, and people's perception of them vary widely, but stopping to consider tastes beyond the big three—sweet, sour, and umami—will help you pick out even more exceptional, interesting tomato flavors.

Another thing to keep in mind is that each tomato has two different flavors: the juicy/jelly part of the fruit that surrounds the seeds and the firmer, drier flesh under the skin. How much this impacts your selection will depend on the type of tomato you are breeding. Cherry tomatoes generally have a lot of juice and little flesh, so the flavor is mostly from the juice. Paste tomatoes are the opposite, with little juice and a lot of flesh. As a breeder, you'll want to take this into account, because if you try to improve the taste of a paste tomato by crossing it with a delicious cherry variety, you might get nowhere if the flesh of the cherry variety is actually fairly bland. When I'm trying a new variety, I like to taste the two parts of the fruit separately, so I have a better idea of how to use it in my breeding.

When it comes to what makes a tomato delicious, texture is another major player and, again, a very personal one. Some people hate cherry tomatoes that pop in their mouth; others can't stand a tomato with a mealy texture. The best way to figure out what you like is to grow a wide range of tomatoes and taste them all. Chew carefully; be aware of their texture in your mouth, and see what you like best. If you love to cook them down into sauces or chop them into salsa, try them that way too. Remember, the flavor fresh and without salt matters only if that is how you like to eat your tomatoes, so taste them the way you usually eat them.

Delicious as they are, tomatoes are, unfortunately, fundamentally prone to fungal disease. They originate in the baking heat and low humidity of Central America, and faced with higher humidity and irrigation, various spots and blights take their toll. Most heirloom varieties, frankly, perform terribly in this respect; luckily, commercial breeders recognize the seriousness of this problem and have been very busy crossing tomatoes with related, more disease-proof species. These other species tend to have small, green, nearly inedible fruit, but careful commercial breeding has successfully isolated the genes for disease resistance and moved them into modern hybrid varieties. But flavor lags far behind—which makes a great opportunity for the home breeder. Pick your favorite tomato for taste, and the most disease-resistant modern variety in your area, and start putting the best of both worlds together.

For gardeners in northern climates, tolerance of cooler temperatures and early fruiting are critical attributes. Where summers stay chilly, tomatoes barely grow, and it takes so long for fruits to ripen, so you barely get a harvest before the first frost arrives. Many catalogs list early varieties that are supposed to be better for northern climates, but the difference is often fairly minor, and flavor is generally wretched. Here again, much work remains to be done. One intriguing option for home gardeners to improve this trait is *Solanum habrochaites*, the most cold tolerant of the wild tomatoes, hailing from very chilly high altitudes in the Andes. It is difficult for a home gardener to track this species down, but interestingly, the rootstock tomatoes being sold in some catalogs for use in grafting are hybrids between *S. habrochaites* and the regular tomato. These plants have a long way to go before they even approach delicious, but they might be a starting point for really great plants for colder climates (and they also boast impressive disease resistance).

Gardeners in hotter parts of the country have the opposite problem: excessive summer heat can stop tomatoes from producing. A few "hot setting" varieties are available, but there is room for improvement, where expanded diversity and flavor are concerned.

Tomatoes, particularly heirloom varieties, come in a wide range of colors. The basic red tomato is green with chlorophyll when immature, and then as it ripens, the chlorophyll breaks down, and a red pigment, lycopene, is produced, causing it to change from green to red, a process very similar, actually, to leaves turning from green to red in the fall. The other color types found in the cultivated tomato are produced by genetic tweaks to that process. Green ripe tomatoes simply fail to go through that step, so though ripe, they keep their chlorophyll and don't produce lycopene. White tomatoes lose their chlorophyll as usual, with maturity, but fail to produce any other pigments. The so-called black tomatoes still produce the lycopene, but their chlorophyll doesn't break down, so the final fruit has both the red and green pigments, making them a brownish color. Yellow tomatoes result from a third class of pigments, carotenoids, which are always present under the regular red, but not visible except in varieties that fail to produce red color at all.

I've heard that different colors are connected to different flavors, but in my experience that isn't true. In my populations, I don't taste any difference between the blacks and the reds.

There are three main growth habits in tomatoes. The first is indeterminate, which is how wild tomatoes grow; each branch can grow indefinitely, forming a large, vining plant that needs staking and continues producing flowers and fruit over a prolonged period of time. Determinate tomatoes are the result of a single recessive gene and grow branches that reach a specific length and then stop; often, these have been bred to produce fruit in a single concentrated burst, though some will continue to produce over a longer period. Most heirloom varieties are indeterminate, while modern breeding favors determinate forms: they are well adapted to mass production (concentrated fruit set lets them all be harvested at one time), and they don't require staking. The third, and still very rare, growth type, produced by another single recessive gene, is the dwarf or "tree" tomato growth habit. These are short, compact plants that (like indeterminates) continue to grow and produce all season, but (like determinates) they don't require staking. Win-win. I predict this last habit will dominate the world of home gardeners in the coming years.

Crossing tomatoes isn't too hard, but it takes a little practice. Since they are self-pollinators, tomatoes need to be emasculated about one day before their small and somewhat delicate flowers would naturally open. Look for buds where the petals are beginning to show yellow but have not yet spread open. With fine-tipped tweezers, pull off the petals, and you'll see the anthers forming a yellow cone around the stigma. Removing the anthers without damaging the stigma can be difficult; plan on giving yourself practice with a few buds before you do it perfectly. My preferred technique is to firmly hold the bud with my off (left) hand, then with the tweezers in my other hand, gently pinch the base of one of the anthers and pull it up and off. Once one anther has been removed, it is easy to gently grab and pull the others off. The critical thing is to leave the stigma unscathed. Some varieties are easier to do this with than others. Generally, the bigger the fruit, the bigger the flowers. I find the flowers of very small cherry tomatoes almost impossible to emasculate, so if possible, I use them as the male parent in my crosses. Beefsteak tomatoes are particularly easy to emasculate because they have a deformed flower structure with many anthers irregularly placed and easy to pull off, unlike the usual tightly formed cone of anthers.

Once your female has been emasculated, take a male flower that has just recently opened and pull off an anther. You won't see any loose, powdery pollen, as on a tomato the pollen is contained within the fold of the anther, to be shaken out by the wind and the buzz of arriving bees. To get the pollen out, hold the anther in your off hand and, with your dominant hand, take the tip of your tweezers (or any other fine-pointed object) and gently run it down the inside of the anther. You'll see a tiny speck of yellowish powder on your tweezer tip. If you don't get any pollen, the flower might be too old and bees might have taken all your pollen, so try another flower. Once you have some pollen on the tweezer tip, just gently dab it on the tip of the stigma. If you want to be extra careful, you can then cover your pollinated flower to prevent bees from bringing in other pollen, but I haven't found it to be necessary. I do often go back the next day and repollinate; it doesn't seem to be necessary, but it makes me feel better, and it is easy enough.

Even a small cherry tomato contains dozens of seeds, so you really need only one successful cross to get what you need. Still, I usually pollinate two or three flowers at a time—a little insurance in case one gets damaged or nibbled by a rabbit or accidentally eaten.

Tomato seeds, as you probably know, come suspended in a thick slime. The best way to deal with this is to squeeze the seeds out into a cup of water. Let that sit in a warm spot for a few days, and the gel around the seeds will ferment and break down. Pour off the water and fermented slime, then put the seeds on a sheet of paper to dry. I used to put them on paper towels, but the seeds stick horribly, and it is almost impossible to get them off the towels once they dry. I now use regular paper (leftover scrap paper, even junk mail), which seems to release the dry seeds much better.

Tomatoes are naturally self-pollinating plants with minimal inbreeding depression, so they are generally inbred to stabilize. In most cases, if you simply leave a flower alone, it will self-pollinate, but depending on the variety and your local bee population, you may get some crossing between different varieties. To ensure a flower self-pollinates, simply cover it before it opens with cloth or parchment paper to keep bees out, and remove once the flower has faded. I never do, and haven't had any problems with my varieties interbreeding, but your mileage may vary, and it is a minimal amount of work to be sure your seeds are what you want them to be.

The history of amateur tomato breeding is long and varied, but there has never been a better time to get into it. More and more great catalogs provide access to varieties from around the world, and modern commercial breeding has created a host of more genetically diverse, disease-resistant varieties just waiting to be integrated into the beautiful and delicious world of heirlooms. It is true: the only thing better than a perfect tomato on a summer's day is one you bred yourself.

WORKING WITH NEW GROUPS

The preceding sections on specific plants don't even begin to scratch the surface of the many fascinating groups of plants you can—and should—try your hand at breeding. So how do you figure out how to breed a new, unfamiliar plant? Every group of plants is different genetically, but the same basic principles apply, and if you ask the right questions, you can quickly find your way into a new group of plants and start making wonderful things from them.

How is it propagated?

If it is from cuttings or division, you can simply make crosses and pick out the individuals you like best. If from seed, you need to find out if it is prone to inbreeding depression and plan accordingly, to either develop inbred lines or maintain larger, more diverse populations to keep inbreeding depression at bay. Outcrossing versus selfing can be the hardest bit of information to track down for a new group of plants, but you can get a good clue simply by looking at the flowers. As a general rule, large showy flowers exist to attract pollinators, which of course are only useful to an outcrossing species, while selfers tend to have smaller, less noticeable flowers, since if they are just going to self-pollinate, why waste the energy showing off for the bees? A good example would be to compare the large vividly colored flowers of scarlet runner beans, an outcrosser, with the very closely related, selfing common beans, which have smaller, much less conspicuous flowers. There are numerous exceptions to this rule, and in the end, the only way to tell for sure how significant inbreeding depression is in a group is to inbreed it for a few generations and see what happens. But looking at the flowers is a start.

How long is the generation time?

This is a practical issue, because to successfully breed something, you need to be able to grow it from seedling to flowering. For some annuals, this can happen in as little as a few months, allowing you to rapidly go from F_1 to F_2 to inbred F_6s, and on to another project. At the opposite extreme, breeding things like fruit trees will force you to wait years for your seedlings to mature so you can evaluate them and move on to the next generation. That isn't to say you can't breed apples, but before you do so, you need to be reasonable about how much space the developing trees are going to take up, and how willing you are to be patient to see the results.

How do I make crosses?

To figure out the mechanics of making a cross, simply grab some flowers and start pulling them apart. A few basic arrangements define virtually all flowers, and once you've seen a few, you'll get pretty good at recognizing the parts. Looking for powdery pollen is the dead giveaway for the male parts of the flower. The female parts will often have a sticky tip, but they are most easily recognized simply by being the bit in the flower that isn't the male part. Try emasculating and pollinating at different stages of development, and see what works best. A surprising number of plants are self-incompatible, which can make your life easier, since you won't have to emasculate them to prevent self-pollination. Try covering a few flowers with something to exclude pollinators and see if you still get seeds. If not, you can probably skip emasculation without worrying about self-pollination. I also like to remove the petals but leave the flower uncovered, to see if pollinators will still visit it. If no seeds set, you don't have to worry.

What do I have to work with?

Spend some time with catalogs and books and find out what diversity there is in the group. Are there obscure varieties with interesting attributes that could be moved into more practical forms? Are there other species that can be included in crosses? Often it can be all but impossible to find out how easily one species will cross with another. The reality, especially with ornamentals, is that most of the possible crosses simply have never been tried, and there is no sure way to predict what crosses will work and which won't before you give it a shot.

You can get a hint from two pieces of information. First, look up the chromosome numbers at tropicos.org/ Project/IPCN. If two species have different chromosome numbers, they will often fail to cross, or be sterile if they do. The other thing to look for is phylogenetic trees. Basically, these are analyses of how genetically similar two species are, arranged in a branching tree pattern, with more closely related species grouped together on the same branch and more distantly related species on other branches. Use Google Scholar and search for genus name and words like "phylogeny" or "phylogenetic," and you'll often turn up academic papers with good information, though they are often (frustratingly, given that the studies behind them are usually funded by tax dollars) behind paywalls unless you are at a university.

In general, the closer they are on a phylogenetic tree, the more likely two species are to successfully hybridize. In general. There are loads of exceptions, and in the end, you just have to try it. And sometimes, if you are really interested in the particular cross, try it again— many, many times. Even if someone else tried and failed, that doesn't mean it won't work in different conditions, or with different varieties of the species. I make a lot of crosses that will probably fail. I don't put a lot of effort into them—just quickly dab some pollen and see what happens. Most times they fail. But every now and then, something works, and the results, when it does, have usually been incredibly fun.

Who else is doing this?

This is one of the most powerful questions you can ask, and in the Internet age, the answer has never been easier to find. For a surprising number of plants, you can find an individual or group of people who are already obsessed with breeding the object of your desire. In my experience, almost all such people are also incredibly generous and having a wonderful time; they will happily share information, plants, ideas, and very valuable experiences about what has and hasn't worked for them. Even if you can't find anyone doing what you want to do, it is worth putting your ideas out there on a blog or forum (or somewhere else Google can find it), so that the next person who gets excited about what you are excited about can run into you and you can share what you have learned with them. One of the best things about gardening and breeding in the Internet age is that you will have the opportunity to make some of the best friends of your life, around the country and around the world.

＊

That's it, really. Find the answers to those questions, and you'll be well on your way to creating something beautiful and unique. Working with more obscure groups can be frustrating, because you have to figure out so much yourself, but it is also the most exciting, because you get to explore truly uncharted territory. It is very unlikely that we're going to see any more really big breakthroughs without a lot of work in highly bred groups like hostas and daylilies. But strike out into a little known group of wildflowers that grow well in your area, or some kind of obscure native American vegetable no one grows any more, and you can very easily make real dramatic changes.

RECOMMENDED READING

Breed Your Own Vegetable Varieties / Carol Deppe
Very in-depth guide to vegetable breeding, including an invaluable list of key breeding information for almost every food crop imaginable.

Breeding Ornamental Plants / Dorothy J. Callaway and M. Brett Callaway, eds.
Detailed information on the breeding of several groups of ornamental plants.

Flower Breeding and Genetics / Neil O. Anderson, ed.
Written for an academic audience but still accessible to the serious amateur, with a series of chapters exploring breeding in various groups of ornamental plants in detail.

Hybrid: The History and Science of Plant Breeding /
Noel Kingsbury
A fascinating look at plant breeding through a cultural and historical lens.

Seed to Seed: Seed Saving and Growing Techniques for Vegetable Gardeners / Suzanne Ashworth
This encyclopedic guide is a great resource for information on almost any vegetable in the world.

USEFUL WEBSITES

alanbishop.proboards.com
The Homegrown Goodness forums, mostly devoted
to edible plants and homesteading, includes many
creative breeders.

daffseek.org
The database of information for breeders of daffodils,
with complete information on pedigrees, descendants,
and fertility of daffodil varieties. A wonderful resource
provided by the American Daffodil Society.

forums2.gardenweb.com/forums/annuals
GardenWeb Annuals forum, devoted to annuals in general, but look for a massive, continually renewed thread
titled "It can be fun to breed your own zinnias" to dis-
cover a vibrant community of zinnia breeders.

HelpMeFind.com
A fundamental resource for the rose breeder. Look up
any variety and find its parentage, descendants, and
information and photos from other rose growers and
breeders. Some features require a very modest subscrip-
tion, but most are free. Has similar databases for clematis
and peonies, though they are not as well developed.

rosehybridizers.org
Website of the Rose Hybridizers Association, home to a
wonderful forum of talented, passionate, and extremely
generous rose breeders.

tomatoville.com
Devoted to everything tomato, the Tomatoville forums
have one of the most active, collaborative communities
of backyard breeders I know.

tropicos.org/Project/IPCN
Index to Plant Chromosome Numbers, a fully searchable,
up-to-date reference created by the Missouri Botanical
Garden; it is always one of my first stops when exploring
a breeding project in a new genus.

PLANT SOURCES

Adaptive Seeds
A unique catalog, skewed toward gardeners who are interested in breeding and adapting varieties to their own garden. A great source for the curious vegetable gardener.

Annie's Annuals and Perennials
Annie's is an amazing nursery; she has an eye for both the unusual and the highly garden-worthy—a terrific combo, and a terrific source of inspiration for the breeder of ornamental plants.

Arrowhead Alpines
Don't be fooled by the name: Arrowhead does indeed carry a stunningly wide range of alpine and rock garden plants, but they also have an impressive selection of other perennials, trees, and shrubs. A top source for gardeners in cold climates.

Baker Creek Heirloom Seeds
The mother of all heirloom vegetable seed catalogs with a mind-boggling range of varieties from around the world. A must for vegetable breeders looking to increase diversity in their projects.

B and T World Seeds
Perhaps the most extensive list of unusual seeds I know. Little to no information in the catalog about each variety, but if I am looking for a specific species, and can't find it, B and T almost always comes through. Based in France, ships internationally.

Chiltern Seeds
One of the widest ranging, most diverse, and interesting ornamental seed catalogs out there. You'll find everything from old classics to stuff you've never heard of, all described with a wonderful touch of British humor. Based in the UK, ships internationally.

J. L. Hudson, Seedsman
A charmingly quirky catalog of ornamentals and some vegetables, full of unexpected varieties and species.

Johnny's Selected Seeds
A terrific company, thanks to extensive trialing and an excellent breeding program. Johnny's offers primarily vegetable varieties that nicely balance qualities like flavor and beauty with health and vigor.

Plant Delights Nursery
One of the great rare plant nurseries, stocking a wide range of ornamental perennials, especially those suited to the heat and humidity of the American South.

Plant World Seeds
A UK-based company that ships internationally, Plant World is home to a wonderful selection of mostly ornamental plants, many of them from their own creative breeding programs, most notably their gorgeous, fragrant columbines.

Seed Savers Exchange
They produce a publicly available catalog, but to really benefit, you need to join the organization, which gives you access to the widest range of vegetable varieties out there, bar none. Many heirlooms, but also many varieties from other backyard breeders.

Territorial Seed Company
Vegetable seeds, especially for the Pacific Northwest.

INDEX

ABOUT THE AUTHOR

Joseph Tychonievich studied horticulture, plant breeding, and genetics at Michigan State University. He has also worked at the Ornamental Plant Germplasm Center, the Chadwick Arboretum Learning Gardens, and for the famed nurseryman Akira Shibamichi in Saitama, Japan. He is currently nursery manager for Arrowhead Alpines.

He has been a repeat guest on American Public Media's food show, *The Splendid Table*, and on garden writer Ken Druse's radio podcast, *Real Dirt*. He blogs at GreensparrowGardens.com.

In the garden, he enjoys growing just about everything. He currently has a 0.6-acre lot, which is slowly being converted into a massive garden. He's especially fond of growing things from seed and, unsurprisingly, is addicted to plant breeding. He's hooked on the thrill of seeing an entirely new plant grow and come into bloom.